Steck-Vaughn

THE COMPLETE EDITION

WORLD GEOGRAPHY AND YOU

Vivian Bernstein

Consultant

Jacquelyn Harrison, Ed. D.
Curriculum & Instruction Administrator for Social Studies
Round Rock Independent School District
Round Rock, Texas

STECK-VAUGHN
ELEMENTARY · SECONDARY · ADULT · LIBRARY

A Harcourt Company

www.steck-vaughn.com

ABOUT THE AUTHOR

Vivian Bernstein is the author of *America's Story*, *World History and You*, *America's History: Land of Liberty*, *American Government*, and *Decisions for Health*. She received her Master of Arts degree from New York University. Bernstein is active with professional organizations in social studies, education, and reading. She gives presentations about content area reading to school faculties and professional groups. Bernstein was a teacher in the New York City Public School System for a number of years.

STAFF CREDITS

Executive Editor: Ellen Northcutt
Design Manager: Rusty Kaim

Photo Editor: Margie Foster
Electronic Production: Jill Klinger

ACKNOWLEDGEMENTS

Cartography: GeoSystems, Inc.
Illustrations: Academy Artworks
Flags: © The Flag Folio
Photography Credits: (KEY: C=Corbis; CB=Corbis-Bettmann; CL= Corel; SS=Superstock)
P. 2 © SS; p. 3 Wyoming Division of Tourism; p. 4 (top) © SS, (bottom) © Didier Doryal/Masterfile; p. 7 (left) © J. A. Kraulis/Masterfile, (right) © SS; p. 9 © Charles Krebs/The Stock Market; p. 10 (top) © Hans Blohm/Masterfile, (bottom) CB; p. 11 (both) © Peter Christopher/Masterfile; p. 12 © Jed Jacobson/AllSport; p. 16 © SS; p. 17 © Ron Sanford/The Stock Market; p. 18 (top) © Lloyd Sutton/Masterfile, (left) © SS, p. 19 (right) © Tom Tracy/The Stock Market, (bottom) U.S. Capitol Office of Photography; p. 20 NASA; p. 25 © SS; p. 26 (left) © Thomas Braise/The Stock Market, (bottom) © Sherman Hines/Masterfile; p. 27 (top) © SS, (right) © Patrice Halley/Gamma Liaison; p. 28 (top) © J. A. Kraulis/Masterfile, (bottom) © Andrew Vaughn/; p. 29 © Roland Weber/Masterfile; p. 34 (both) AP/Wide World; p. 35 (top) © SS, (bottom) © Mark Tomalty/Masterfile; p. 36 (top) © SS, (bottom) © Lou Jr. Jacobs/Gamma Liaison; p. 39 (left) © Bo Vince Street/The Stock Market, (right) © W. Bayer/Bruce Coleman, Inc.; p. 42 (top) © Alpamayo John Phelan/DDB Stock Photo, (bottom) © Keith Gunner/Bruce Coleman, Inc.; p. 43 (right) © J. P. Courau/DDB Stock Photo, (bottom) © F. Erize/Bruce Coleman, Inc.; p. 44 (top) © Paulo Fridman/International Stock, (bottom) © M. Joly/DDB Stock Photo; p. 48 The Granger Collection; p. 49 (top) © Bryon Augustin/DDB Stock Photo, (bottom) © D. Donne Bryant/DDB Stock Photo; p. 50 (top) © D. Donne Bryant/DDB Stock Photo, (bottom) © Jerry Cooke/Photo Researchers; p. 51 (top) © D. Donne Bryant/DDB Stock Photo, (right) © Jean-Marc Giboux/Gamma Liaison; p. 52 © Russell/Bruce Coleman, Inc.; p. 57 © Jaris Burger/Bruce Coleman, Inc.; p. 58 (top) © Claude Urraca/Sygma, (bottom) © Silvio/Gamma Liaison; p. 59 (top) © J. C. Carton/Bruce Coleman, Inc., (right) © Inga Spence/DDB Stock Photo; p. 60 © Chuck Mason/International Stock; p. 65 © Gary Williams/Gamma Liaison; p. 66 (left) © Joan Laconetti/Bruce Coleman, Inc., (bottom) © Sullivan & Rogers/Bruce Coleman, Inc.; p. 67 (top) © Jase Azel/The Stock Market, (right) © SS; p. 68 © Buu-Hires/Gamma Liaison; p. 73 © Jonathan Kirn/Gamma Liaison; p. 74 (top) © Paulo Fridman/International Stock, (bottom) © Erwin & Peggy/Bruce Coleman, Inc.; p. 75 (top) © Editora Abril/Gamma Liaison, (bottom) © Gamma Liaison; p. 76 (top) © John Chiasson/Gamma Liaison, (bottom) © Editora Abril/Gamma Liaison; p. 81 © Bruce Coleman, Inc.; p. 82 © Alejandro Balaguer/Sygma; p. 83 (top) © Alejandro Balaguer/Sygma, (bottom) © Jacques M Chenet/Gamma Liaison; p. 84 (top) © J. C. Carton/Bruce Coleman, Inc., (bottom) © Robin Schwartz/International Stock; p. 89 (top) © David Madizon/Bruce Coleman, Inc., (bottom) © Bill Wrenn/International Stock Photo; p. 90 (top) © Francisco Erize/Bruce Coleman, Inc., (bottom) © George Ancona/International Stock; p. 91 (top) © Norman Owen Tomalin/Bruce Coleman, Inc., (bottom) © Buddy Mays/International Stock; p. 92 (top) © Bleibtreu/Sygma, (bottom) © Roberto Arakaki/International Stock, (bottom) © Hilary Wilkes/International Stock; p. 97 © Gian Luigi Scarifiotti/International Stock; p. 98 © SS; p. 99 (top) © J. Messerschmidt/Bruce Coleman, Inc., (bottom) © Phototheque SGM/International Stock; p. 100 (top) © SS, (bottom) © Joachim Messerschmidt/Bruce Coleman, Inc.; p. 104 © Hilary Wilkes/International Stock; p. 105 (top) © Bruce Coleman, Inc., (right) © Richard Folwell/Photo Researchers; p. 106 (both) © C; p. 107 (top) © Gamma Liaison, (right) © SS; p. 113 © SS; p. 114 (left) © John Elk III/Bruce Coleman, Inc., (bottom) © SS; p. 115 (both) © SS; p. 116 (top) © SS, (bottom) © R. T. Nowitz/Photo Researchers; p. 122 (top) © Bouvet/Hires/Merillon/Piel/Gamma Liaison, (left) © C; p. 123 (right) © Anthony Suau/Gamma Liaison, (bottom) © SS; p. 124 (both) © C; p. 125 (top) © David R. Frazier/Photo Researchers, (bottom) © C; p. 130 © Chazot/Explorer/Photo Researchers; p. 131 © H. P. Merten/The Stock Market; p. 132 (top) © C, (left) © Alvero De Leiva/Gamma Liaison; p. 133 (both) © C; p. 138 © CB; p. 139 (right) © Reuters/Pawel Kapczynski/Archive Photos, (bottom) © Simon Fraser/Science Photo Library/Photo Researchers; p. 140 © Sygma; p. 143 (left) © Loubat/Explorer/Photo Researchers, (right) © Vanier/Explorer/Photo Researchers; p. 145 The Granger Collection; p. 146 © Georges Merillon/Gamma Liaison; p. 147 (top) © Brent Winebrenner/International Stock, (bottom) © C; p. 148 (top) © Bill Bachmann/Photo Researchers, (left) © C; p. 152 © J.P. Laffont/Sygma, 153 (top) © Brossard/Explorer/Photo Researchers, (bottom) © Michael Philip Mannheim/International Stock; p. 154 (top) © Kok/Gamma Liaison, (left) © Laski Diffusion/Gamma Liaison; p. 155 (top) © Boutin/Explorer/Photo Researchers, (bottom) © Archive Photos; p. 160 © Wolfgang Kaehler/Gamma Liaison; p. 161 (top) © C, (right) © Ellen Rooney/International Stock; p. 162 (top) © Photo Researchers, (bottom) © James D. Wilson/Gamma Liaison; p. 163 (top) © Osvald/Gamma Liaison, (right) © Francis Apesteguy/Gamma Liaison; p. 164 (top) © CB, (left) © Heidi Bradner/Gamma Liaison; p. 165 © L. Veisman/Bruce Coleman, Inc.; p. 171 (top) © Buddy Mays/International Stock, (right) © Bruce Brander/Photo Researchers; p. 172 © Novosti/Photo Researchers; p. 173 (top) © Jeff Greenberg/Photo Researchers, (bottom) Gamma Liason; p. 178 © Andrew Reid/Gamma Liaison; p. 179 © Art Zamur/Gamma Liaison; p. 180 (top) © Peterson/Gamma Liaison, (bottom) C; p. 185 (top) C, (right) © Chad Ehlers/International Stock; p. 186 (top) © CB, (left) © V. Leloup/Gamma Liaison; p. 187 (top) © Gilles Saussier/Gamma Liaison, (bottom) © Reuters/David Brauchli/Archive Photos. Continued on page 406.

CONTENTS

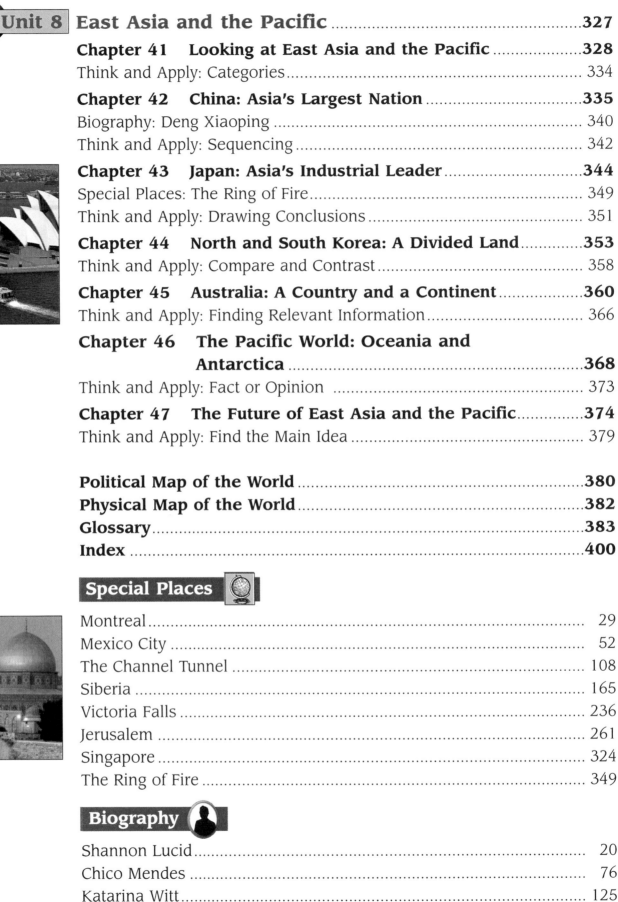

Special Places

Biography

Skill Builder

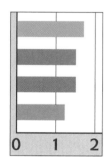

Maps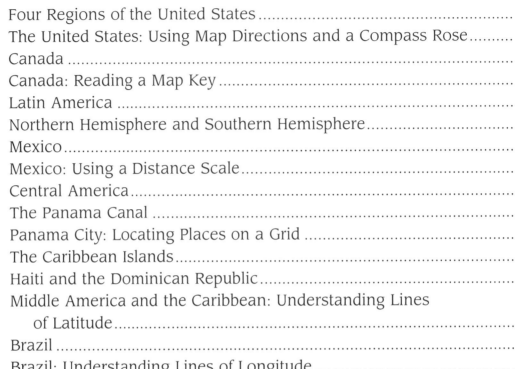

Charts, Graphs, and Diagrams

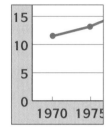

Getting Started in Geography

Where Can You Find?
Where can you find the place where denim jeans were first worn?

Think About As You Read

1. What can you learn by studying world geography?
2. What are the five themes of geography?
3. What are Earth's oceans and continents?

New Words

- geography
- culture
- religion
- themes
- location
- place
- human/environment interaction
- movement
- region
- latitude
- longitude
- landforms
- plateaus
- continents

People and Places

- Brasília
- Brazil
- Atlantic Ocean
- Anápolis
- Pacific Ocean
- Indian Ocean
- Arctic Ocean
- Asia
- Africa
- North America
- South America
- Antarctica
- Europe
- Australia

Where are the longest rivers? How do people live in a desert? Why do people move from one place to another? These are just a few of the questions you will answer as you study world geography.

What Is the Study of Geography?

Geography is the study of the planet Earth. As you study world geography, you will learn what the land looks like in different parts of the world. You will learn how people live in different places. You will study how people have changed the places where they live.

You will also study the **culture** of people in different countries. Culture is the ideas, art, and way of life of a group of people. When you study about culture, you learn about food, clothing, sports, customs, and language of a group of people. **Religion** is also part of culture. Religion is the different ways people believe and pray to a god or to many gods. Culture

Brazil's leaders meet in these buildings in Brasília. This capital city is only about 50 years old.

Brasília, the capital of Brazil

is what makes one group of people different from another.

The Five Themes of Geography

We use five **themes**, or important ideas, to help us study different places. These five themes help us answer important questions about different parts of the world. The five themes of geography are: **location**, **place**, **human/environment interaction**, **movement**, and **region**. You will use the five themes again and again as you study world geography.

Let's look at each of the themes:

Location: This theme helps us answer the question, "Where is the place?" Sometimes the answer to this question describes what a place is near. The location of a place might be near a river, a mountain, or a city. For example, Brasília, the capital city of Brazil, is 600 miles from the Atlantic Ocean. It is near a smaller city called Anápolis. Think about the place where you live. How would you describe its location?

Sometimes we need to find the exact spot on a map or a globe where a place can be found. To find the exact location of a place, we use lines of **latitude** and lines of **longitude**. Lines of latitude are lines on a map that go from east to west. Lines of longitude are lines that go from north to south. Every place on Earth has its own latitude and longitude. The exact location of Brasília is 16° south latitude and 48° west longitude. No other place on Earth is at that same location.

Place: This theme helps us answer the question, "What makes this place different from all other places on Earth?" A place may be different because of its culture. Brasília is a special place because it was built to be Brazil's new capital city almost fifty years ago. All of the city's buildings look modern. Also, it was built in an area that had few people. Today Brasília is a big city with almost two million people. For these reasons, there is no other place like Brasília.

A place can also be different because of its **landforms**. A landform is the shape of an area on Earth's surface. Mountains, hills, plains, and **plateaus** are the four main kinds of landforms. Plateaus are high, flat land. Think about where you live. What makes it different from every other place?

Mountains are a kind of landform.

Human/Environment Interaction: This theme answers the questions, "How do people use and work with the place? How do they change the place?" This theme tells how people can change a place. It also tells how people are changed by a place. By building Brasília, the people of Brazil changed that part of their country. It changed from a quiet area with few people to a large city with many government workers. Before the city was built, there was little transportation in that area. New roads, highways, and an airport were built to connect the new city to other parts of Brazil.

This theme also tells us how people change the way they live and work because of the place. For example,

LANDFORMS

Plateau Mountains Hills Plain

Shown here are four types of landforms. Which landforms can be found where you live?

People often change the place where they live. These trees were cut down to make a road to Brasília.

This ship will move goods from one country to another.

people who live near an ocean or a lake often earn a living by fishing. In places where gold and silver can be found, many people become miners. Think about where you live. How have people used, changed, and worked with your community's land and water?

Movement: This theme helps us answer the question, "How and why do people, goods, and ideas move from place to place?" Every year millions of people move from one place to another. After Brasília was built, government leaders moved to the new city. Then many other people moved there. They started businesses such as restaurants and stores. More people moved there, and the city grew larger.

Trade is the movement of goods from and to different places. Cars and clothes are made in many different countries. Ships and trains bring them to our country. Goods made in our country are shipped to other countries, too.

Culture also moves from one country to another. Americans were the first to wear denim jeans. Today people around the world wear the same kind of jeans that Americans wear. Think about where you live. Which people and goods have moved in and out of your community?

Region: To study Earth, people divide it into areas called regions. A region is an area in which the people or places share something important. Brasília is in the region that has Brazil's government. The city has government buildings and homes for government leaders. Often a region is described by its landforms. For example, a region may be covered with tall mountains. Places in a region may share the same climate. Sometimes the people in a region share the same culture. In what region is your community located?

Oceans and Continents

As you study world geography, you will learn about Earth's oceans and **continents**. A continent is a large body of land. Most of Earth is covered with water. Much of the water is in four large oceans. The oceans

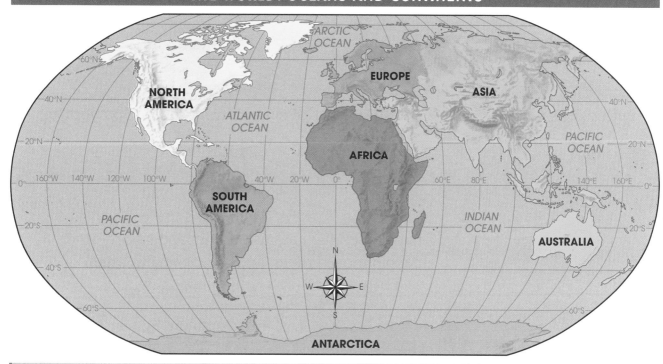

Most of Earth is covered with water. Most of Earth's land is found on seven continents. What are the names of the largest ocean and the largest continent?

in order of size from largest to smallest are the Pacific Ocean, the Atlantic Ocean, the Indian Ocean, and the Arctic Ocean.

Most of Earth's land is found on seven continents. The seven continents from largest to smallest are Asia, Africa, North America, South America, Antarctica, Europe, and Australia. On which of these continents is your home?

As you study world geography, you will learn about different regions and countries. Think about the ways each place is different or the same as your community.

Chapter Main Ideas

1. Geography is the study of the planet Earth.
2. The five themes of geography are location, place, human/environment interaction, movement, and region.
3. Earth has four oceans and seven continents.

◆ Vocabulary

Match Up   Number your paper from 1 to 5. Finish the sentences in Group A with words from Group B. Write the letter of each correct answer on your paper.

Group A

1. A _____ is the ideas, art, and way of life of a group of people.

2. The geography theme of _____ tells where to find a place.

3. A _____ is a large body of land.

4. A _____ is an area that shares something important.

5. The theme of _____ tells how people use and change a place.

Group B

A. region

B. culture

C. human/environment interaction

D. continent

E. location

◆ Read and Remember

Write the Answer   Number your paper from 1 to 5. Write one or more sentences to answer each question.

1. What are the five themes of geography?

2. How is Brasília different from all other places?

3. How has movement changed the area of Brasília?

4. What are the names of the four oceans?

5. What are the four main types of landforms?

◆ Journal Writing

Write a paragraph in your journal that tells the location of the place where you live. What makes this place different from other places you know?

The United States and Canada

Rocky Mountains

Niagara Falls

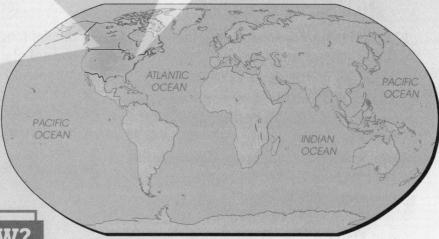

ATLANTIC OCEAN

PACIFIC OCEAN

PACIFIC OCEAN

INDIAN OCEAN

DID YOU KNOW?

▲ The name of Canada comes from *Kanata*, an American Indian word which means "village."

▲ Almost half of the fruits and vegetables grown in the United States come from California.

▲ It rains up to 350 days per year at Mt. Waialeale, Hawaii.

▲ In 1969 a dam was built to stop the water from flowing over the American side of Niagara Falls for a short time.

WRITE A TRAVELOGUE

Imagine you are a world traveler. You will be visiting the places described in this first unit. Keep a travelogue or a journal about your trip. Before reading Unit 1, write a paragraph about the places in the United States and Canada you would want to visit. After reading Unit 1, write a paragraph about two interesting places in these countries.

THEME: PLACE

CHAPTER 1

Two Nations of North America

Where Can You Find?
Where can you find the tallest place in North America?

Think About As You Read

1. What landforms are found in the United States and Canada?
2. Compare the climates of the United States and Canada.
3. How are the United States and Canada alike and different?

New Words

- official languages
- coastal plain
- mountain chain
- climates
- immigrants
- freedom of religion
- democracies
- urban
- suburbs
- industrial nations
- standard of living
- developed nations
- unguarded border

People and Places

- United States
- Canada
- Rocky Mountains
- Mount McKinley
- Alaska
- Great Lakes
- Arctic
- Hawaii
- Native Americans
- American Indians
- Spain
- France
- Great Britain
- Spanish
- French
- British

Two friends are visiting a large shopping mall in North America. The sound of rock music can be heard in many of the stores. Some of these stores sell jeans and sweatshirts. Other stores sell toys, telephones, or books. The two friends buy hamburgers for lunch.

Can you tell if this mall is in the United States or in Canada? It may be difficult to tell because both nations are alike in many ways. The signs in the mall can give you the answer. The signs in this mall are in both English and French. Both English and French are the **official languages** of Canada. The United States does not have an official language. Most people in the United States speak English as their main language.

Understanding the Land and the Climate

The United States and Canada are the two largest countries in North America. Canada is the world's second largest country in size. The United States is

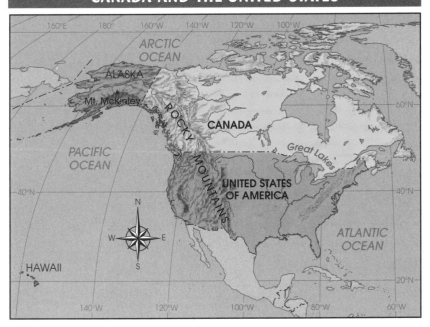

Canada and the United States share some of the same landforms. What mountain chain can be found in both countries?

the fourth largest country in the world. Find the United States and Canada on the map above.

The United States and Canada are between two large oceans. The Atlantic Ocean is to the east of both nations. The Pacific Ocean is to the west. The icy Arctic Ocean is north of Canada. The Atlantic and Pacific oceans help both nations. Both countries use the oceans for fishing. They use the oceans for trading with countries around the world.

The United States and Canada have the same kind of landforms. In the east both have a **coastal plain** along the Atlantic Ocean. A coastal plain is flat land near an ocean. Low mountains are to the west of the coastal plain. The tall Rocky Mountains are in the west. This **mountain chain** is 3,000 miles long. A mountain chain is a long group of mountains. One of these mountains, Mount McKinley in Alaska, is the tallest mountain in North America. Large plains are between the Rocky Mountains and the mountains in the east. These plains have good farmland.

Both countries have many large lakes. Four of the five Great Lakes are shared by the United States and Canada. The Great Lakes are among the largest lakes in the world.

Mount McKinley, Alaska

The Arctic in northern Canada is a very cold place to live.

Immigrants arrive in the United States.

The two countries have many different kinds of **climates**. Northern Canada and northern Alaska are in a region called the Arctic. This region is always very cold. Most of southern Canada has warm summers and very cold winters.

The United States has many different climates. Many northern states have long, cold winters and hot summers. Hawaii and some western and southern states have warm weather throughout the year.

History, People, and Government

For thousands of years, Native Americans, or American Indians, lived in many parts of North and South America. About 500 years ago, people from Europe began to settle in North and South America. Explorers from Spain, France, and Great Britain claimed large parts of North America. Spanish people settled in what is now the southern and western parts of the United States. Much of what is now Canada was settled by French people. British people settled in the eastern part of what is now the United States.

The French and the British fought against each other to rule more land in North America. They fought several wars. Finally, in 1763 the British won. After that, Great Britain ruled Canada. Many British people moved to Canada. Canada belonged to Great Britain until 1931. Today Canada rules itself.

Great Britain had ruled the eastern part of what is now the United States. In 1776 Americans decided to be free from Great Britain's rule. They fought and won a war for freedom. They called their new nation the United States of America.

For hundreds of years, **immigrants** have settled in the United States and in Canada. People from every part of the world now live in these countries. People in both countries have a lot of freedom. All people have the freedom to speak and to write as they wish. There is **freedom of religion** for all. This means people can pray any way they choose.

The United States and Canada are **democracies**. In a democracy, people vote for their leaders. They

Most people in Canada and in the United States live in urban areas. This city is in Canada.

also vote for their lawmakers. But the two countries are two different types of democracies. They have different types of governments. They have different laws and different kinds of leaders. You will read about their governments later in this unit.

Living in the United States and Canada

The United States and Canada are both large nations. However, the United States has ten times more people. In both nations most people live in cities, or **urban** areas. Many people live in areas close to the cities called **suburbs**.

In both nations only a small part of the people are farmers. But these farmers grow far more food than the people of their countries need. Both countries sell food to people in other countries.

The United States and Canada are **industrial nations**. An industrial nation has many factories. Both countries make cars, computers, and many other products. Most people in the United States and Canada have a high **standard of living**. The standard of living measures how well people live in a region. Both countries have plenty of businesses, schools, and hospitals. In both countries most people can afford to own cars, refrigerators, and telephones. A country that has few people who can afford to own these types of goods has a low standard of living.

Car parts are made in this factory in Canada.

The United States and Canada are friendly neighbors. Teams from the United States and Canada play each other in baseball, hockey, basketball, and other sports.

Both the United States and Canada are called **developed nations**. Developed nations are industrial nations with a high standard of living.

Both countries have a lot of trade with other nations around the world. The United States has more trade with Canada than with any other nation. Canada has more trade with the United States than with any other country.

There is real friendship between the two large nations of North America. The two countries share the world's longest **unguarded border**. There are no soldiers along an unguarded border. In the next chapters, you will learn more about the United States and Canada.

Chapter Main Ideas

1. The United States and Canada have the same kind of landforms. There are many different climates in both countries.
2. The United States and Canada are democracies. But they have different types of governments.
3. The United States and Canada are developed countries. Both are industrial nations with a high standard of living.

◆ Vocabulary

Finish Up   Number your paper from 1 to 6. Choose the word or words in dark print that best complete each sentence. Write the word or words on your paper.

developed nation   standard of living   democracy
suburbs   coastal plain   freedom of religion

1. A country that has people that choose their own leaders and write their own laws is a _____.

2. A _____ has many factories, banks, and stores.

3. When a nation has a high _____, most people can afford to own cars, televisions, and telephones.

4. Communities near cities are called _____.

5. Flat land near an ocean is a _____.

6. People with _____ can pray any way they choose.

◆ Read and Remember

Matching   Each item in Group B tells about an item in Group A. Number your paper from 1 to 5. Write the letter of each item in Group B on your paper next to the correct number.

Group A

1. Hawaii

2. industrial nation

3. Arctic

4. French and English

5. Great Britain

Group B

A. This type of nation makes many cars, planes, and computers in its factories.

B. This state in the United States is always warm.

C. This region in northern Canada is always very cold.

D. This nation once ruled Canada.

E. These are Canada's official languages.

Compare and Contrast   Copy the Venn diagram shown below on your paper. Then read each phrase below. Decide whether it tells about Canada or the United States. If it tells about either nation, write the number of the phrase in the correct part of the Venn diagram. If the phrase tells about both nations, write the number of the phrase in the center of the diagram on your paper.

1. industrial nation

2. two official languages

3. many immigrants

4. fought Great Britain for freedom

5. warm climate throughout the year in south

6. democracy

7. second largest country in the world

8. no official language

9. an unguarded border

10. settled by the French

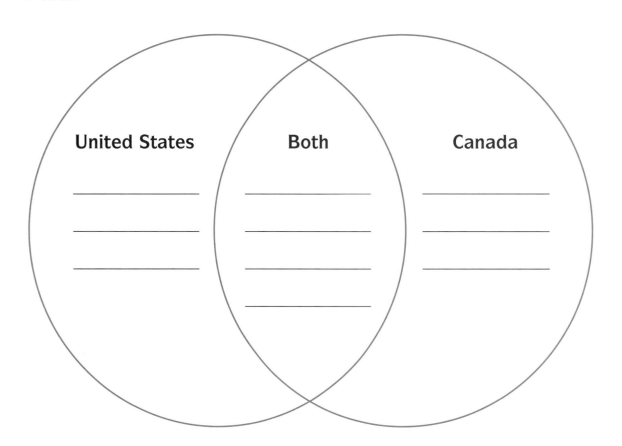

United States   Both   Canada

◆ **Journal Writing**

There are many ways that the United States and Canada are alike and different. Write a paragraph in your journal that tells at least one way the two countries are alike. Then tell one way they are different.

The United States: Land of Variety

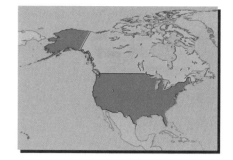

Where Can You Find?
Where can you find the hottest place in the United States?

Think About As You Read

1. How are the four regions of the United States different from each other?
2. What are some natural resources of the United States?
3. What are some problems of the United States?

New Words

- ◆ variety
- ◆ citizens
- ◆ natural resources
- ◆ population density
- ◆ cash crops
- ◆ technology
- ◆ service jobs

People and Places

- ◆ Puerto Rico
- ◆ New York City
- ◆ Los Angeles
- ◆ Chicago
- ◆ Washington, D.C.
- ◆ Northeast
- ◆ Midwest
- ◆ Great Plains
- ◆ Texas
- ◆ Oklahoma
- ◆ Mississippi River
- ◆ Gulf of Mexico
- ◆ Death Valley
- ◆ Africans
- ◆ African Americans
- ◆ Shannon Lucid

What do you want to do during your next winter vacation? Do you want to go swimming at a warm beach? Do you want to ski down a mountain? Do you want to see a play in a big city? In the United States, you could do all of these things and many more. The United States is a land of great **variety**.

Many States and Cities

In Chapter 1 you read that in 1776 Americans decided to be free from Great Britain. When the United States first became a free nation, it had only 13 states. Soon more states became part of the country. Today the United States is made up of 50 states. In 1959 Alaska became the forty-ninth state. Later that year Hawaii became the fiftieth state. These two states are far from all the other states. Alaska is north and west of Canada. Hawaii is a group of islands in the Pacific Ocean. The people in all 50 states are **citizens** of the United States. A citizen is a member of a country.

The United States is made up of 50 states. Which two states are far from all the other states?

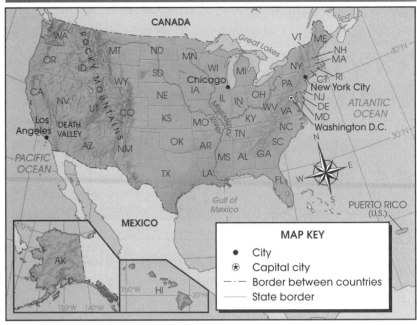

THE UNITED STATES OF AMERICA

Flag of the United States of America

The Capitol in Washington, D.C.

Some parts of the United States are not states. One of these places is Puerto Rico. This island belongs to the United States. Puerto Rico is not a state. But the people of Puerto Rico are citizens of the United States.

There are many large cities throughout the United States. New York City, Los Angeles, and Chicago are the largest cities in the United States. Millions of people live in these cities. Millions more live in the suburbs close to the nation's cities. Washington, D.C., is the capital of the United States. The leaders of the United States work in Washington, D.C.

Regions of the United States

We can divide the United States into regions in order to study it. Let us look at four large regions—the Northeast, the Midwest, the South, and the West.

The first region is the Northeast. Its landforms are coastal plains and low mountains. It has many harbors. The Northeast region has fewer **natural resources** than other regions. Natural resources are things we get from Earth. Coal, fish, and forests are the most important natural resources of the Northeast.

The Northeast has many large cities. This region has the nation's highest **population density**. This means that more people live closer together in this region than in any other part of the country. Many people work in factory jobs. There are also many people who work at fishing and trading.

The second region is the Midwest. It is near the Great Lakes. This region has coal and iron. There are many factories in the cities next to the Great Lakes. The factories use coal and iron to make cars and machines. Ships on the Great Lakes carry these products to many places.

The Midwest has huge flat plains. Some of these plains are called the Great Plains. Wheat is grown in this region. Bread, cereal, and flour are made from wheat. This region also has many dairy farms.

The South is the third region. Many people are moving to the South from other parts of the country. This region has a large coastal plain. It also has low mountains. Its warm climate is good for farming. Farmers grow **cash crops**, or crops they can sell. Some cash crops are cotton, rice, and oranges.

Oil is an important natural resource in states such as Texas and Oklahoma. Oil is made into many products such as gasoline for cars and planes.

The Mississippi River runs through many states in the Midwest and the South. The Mississippi starts in the Midwest in the north near Canada. The river is more than 2,000 miles long. It ends in the South at the Gulf of Mexico. Many rivers flow into the Mississippi.

The fourth region is the West. This is a region with many contrasts. The tall Rocky Mountains are in this region. The lowest place in all of North and South America is also in this region. It is Death Valley. Death Valley is the hottest place in the United States. The West has plains and valleys with good farmland. Many fruits and vegetables are grown in the valleys of the West. But there are also dry deserts in the West.

The West has many natural resources. Alaska has large amounts of oil. Many workers cut down trees in the forests in the West. Other people in the region work at mining or fishing.

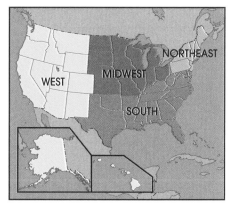

Four Regions of the United States

Death Valley

Americans use new technology to make many products.

How Do Americans Earn a Living?

There are many ways for Americans to earn a living. The land, the climate, and the natural resources of a region help people decide what kind of work to do. Farmers are a small part of the population. Many people work at factory jobs. Factories use new **technology** to make many new products. Technology is the inventions that improve the ways we live and work. Each year Americans build better cars, computers, and machines.

Most Americans now work at **service jobs**. Service jobs are jobs in which people help other people. People who work in schools, hospitals, banks, and stores are doing service jobs. People who work for the government have service jobs.

People who work as teachers have service jobs.

People of the United States

The United States has always been a country with people from many nations. The first Americans, the Indians, continue to live in many parts of the country. People from Europe were the first to move to America from other lands. Now most immigrants come from Asia, South America, and parts of North America.

In 1619 the first Africans were brought to America. Later more and more Africans were forced to be slaves in America. Slavery ended more than 130 years ago. Today more than 30 million African Americans live in the United States.

Today there are many different groups of people in the United States. They follow different religions. Some of them speak different languages. Each year thousands of new immigrants move to America. Some come to get better jobs. Many others come to have more freedom.

Looking at the Future

The United States is a strong, rich country. It is strong because it has a large army. It is also strong because it has good laws and good leaders. The government of the United States is a democracy. The leader of the country is the President. People vote every four years for a President. The people also vote for their lawmakers. The lawmakers meet in a group called Congress.

The United States is a rich country because it has so many natural resources, businesses, and factories. Excellent transportation has helped the country become rich. There are good highways, airports, and railroads. People and goods are able to move to all parts of the country.

Today the country does not have enough natural resources. Americans must now buy a lot of oil from other countries. Americans must work harder to save their own natural resources for the future.

The United States has very good highways. These roads help move goods around the country.

Lawmakers meet in a group called Congress. Here the President speaks to the people in Congress.

The United States is a rich country. But millions of Americans are poor. Some people do not have enough food. Other people do not have homes. Americans must find new ways to help the poor.

The United States tries to help other countries. It gives money and food to other countries. Americans work to bring peace to places that have wars. In the future the United States will continue to help other countries. And it will continue to be a land of variety.

Chapter Main Ideas

1. The United States is a land of variety. It has many different landforms, natural resources, and jobs.
2. The United States has four large regions. They are the Northeast, Midwest, South, and West.
3. People from many countries live in the United States. Many people move to the United States to have more freedom.

BIOGRAPHY

Shannon Lucid (Born 1943)

As a young girl, Shannon Lucid had a dream. She wanted to explore space. Through hard work her dream came true.

After she finished college, Lucid became a pilot. Later she studied to be an astronaut. She flew on several space trips. Then, in 1996 Lucid made history. She spent 188 days in space. During that time she worked for the United States on the Russian space station *Mir*. While in space she did many science experiments. She also exercised to keep her body strong. During the months in space, she did not see her family or friends. After six months Lucid returned to Earth. Americans everywhere were proud of Shannon Lucid. She had spent more time in space than any other American. From her work Americans have learned more about space travel.

Journal Writing
Write a paragraph in your journal that tells why Shannon Lucid is a hero to many Americans.

◆ Vocabulary

Finish the Paragraph   Number your paper from 1 to 6. Use the words in dark print to finish the paragraph below. On your paper write the words you choose.

cash crops	**natural resources**	**variety**
service jobs	**population density**	**technology**

The United States is a land of ___**1**___ because it has many jobs, landforms, and natural resources. The Northeast has a high ___**2**___ because many people live close together in this region. In the South farmers grow ___**3**___, or crops they can sell. They grow cotton, rice, and oranges. The West has many ___**4**___, or things we get from Earth. Most Americans work at ___**5**___ where they help other people. Many people work in factories. These factories use modern ___**6**___, inventions that improve the way people work.

◆ Read and Remember

Complete the Chart   Copy the chart shown below on your paper. Use the facts from the chapter to complete the chart. You can read the chapter again to find facts you do not remember.

Regions of the United States

	Northeast	Midwest	South	West
What are the landforms?				
What are the natural resources?				
What work do people do?				

Matching   Each item in Group B tells about an item in Group A. On your paper write the letter of each item in Group B next to the correct number.

Group A

1. Washington, D.C.

2. Shannon Lucid

3. Puerto Rico

4. Oklahoma

5. Great Plains

6. Mississippi River

Group B

A. This city is the capital of the United States.

B. The people of this island are citizens of the United States.

C. Oil is an important resource from this state.

D. This astronaut was in space for 188 days.

E. This waterway flows south from the Midwest near Canada. It ends at the Gulf of Mexico.

F. This huge flat area in the Midwest is important for growing wheat.

◆ Think and Apply

Categories   Number your paper from 1 to 4. Read the words in each group. Decide how they are alike. Find the best title for each group from the words in dark print. Write the title on your paper next to the correct number.

American Democracy
Natural Resources

Landforms in the United States
Largest Cities in the United States

1. vote for President
   vote for lawmakers
   freedom of religion

2. New York City
   Los Angeles
   Chicago

3. forests and fish
   oil
   gold, coal, silver, and iron

4. Rocky Mountains
   Great Plains
   Death Valley

◆ Journal Writing

You read about four regions in this chapter. Write a paragraph in your journal that tells which region you might want to live in. Give two or more reasons why you chose this region.

Using Map Directions and a Compass Rose

There are four main directions. They are **north**, **south**, **east**, and **west**. There are also four in-between directions. They are **northeast**, **southeast**, **northwest**, **southwest**. A **compass rose** is used to show directions on a map. Sometimes the directions are shortened to **N**, **S**, **E**, and **W**. The in-between directions are shortened to **NE**, **SE**, **NW**, and **SW**.

A. Copy the compass rose shown below on your paper. Write the four main directions and the four in-between directions in the correct places on your compass rose.

B. Look at the map of the United States below. Number your paper from 1 to 8. Choose an answer from the box to finish each sentence. Use the compass rose to help you. Write the correct answers on your paper.

northwest	northeast	southwest
west	north	east
south	southeast	

1. The state of Maine is _____ of Louisiana.

2. The Gulf of Mexico is _____ of Louisiana.

3. The Great Lakes are _____ of Chicago.

4. The Pacific Ocean is _____ of California.

5. Oregon is _____ of New Mexico.

6. Chicago is _____ of Oregon.

7. New Mexico is _____ of Chicago.

8. Maine is _____ of the Great Lakes.

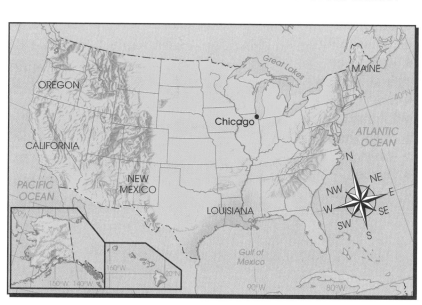

CHAPTER 3

Canada: A Giant Land of Few People

Where Can You Find?
Where can you find the oldest city in Canada?

Think About As You Read

1. Why do most Canadians live in the southern part of Canada?
2. What are some of Canada's regions?
3. What kind of government does Canada have?

New Words

- provinces
- territories
- permafrost
- logging
- newsprint
- lowlands
- waterways
- British Commonwealth
- prime minister
- Parliament

People and Places

- Canadian Shield
- Hudson Bay
- St. Lawrence Lowlands
- St. Lawrence River
- Toronto
- Montreal
- Ottawa
- Quebec City
- Vancouver
- First Nations
- Inuit
- Eskimos
- Northwest Territories
- Nunavut
- Quebec
- French Canadians

Perhaps one summer you will visit northern Canada. Although it may be summer, the air will feel cold. People will be wearing warm jackets. You might want to visit a big city. You will not find one. Few people live in this part of Canada. Most Canadians live in the warmer southern part of Canada.

Canada's Land, Climate, and Natural Resources

In many ways Canada seems like the United States. There are coastal plains and low mountains in the east. There are the tall Rocky Mountains in the west. There are large flat plains between the mountains. But Canada is different.

Canada is a very big country. It has ten states called **provinces**. Each province has its own government. To the north of the provinces are two **territories**. Many islands in the Arctic Ocean are part of these territories. Fewer than 100,000 people live in these two northern territories.

CANADA

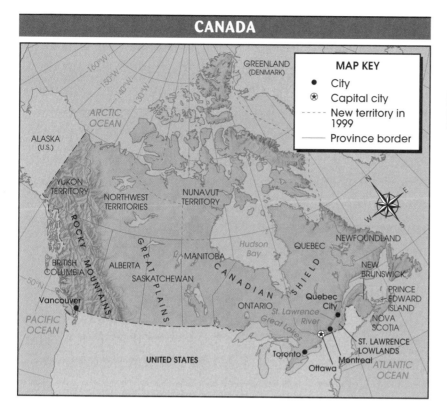

MAP KEY
- • City
- ⊛ Capital city
- ----- New territory in 1999
- —— Province border

Canada has ten provinces. In 1999 Canada will have a third territory. What is the name of the new territory?

Canada's flag

Canada has a small population. Only 28 million people live in this giant country. Most Canadians live within 200 miles of Canada's southern border. This area is warmer than northern Canada. In the summer the temperature in southern Canada is usually about 70°F. But in northern Canada it can be below 40°F. In most of Canada, the temperature during the winter is below 0°F.

The Arctic is the coldest part of northern Canada. Most of the year the ground is frozen. For a few weeks during the summer, the top layer of soil is not frozen. A few small plants can grow in this wet, muddy soil. But there are no trees here. Below the top layer of soil the ground remains frozen. **Permafrost** is the layer of soil that is always frozen. Crops cannot be grown where there is permafrost. But other parts of Canada have very good soil for growing crops.

Canada has many natural resources. Thick forests cover almost half of the country. Canada has a lot of oil and coal. It has gold, silver, iron, and other metals. There is also plenty of waterpower that is used to make electricity.

Parts of northern Canada are covered with snow during most of the year.

25

What Are Some of Canada's Regions?

The Canadian Shield is a large region that curves around Hudson Bay. Find the Canadian Shield on the map on page 25. The Canadian Shield covers almost half of Canada. Few people live in this cold region. But the region is very important because it has many forests and natural resources. Many people in this region work at **logging**. Logging means cutting down trees in the forests. One important forest product is **newsprint**. Newsprint is the paper used for making newspapers. Canada makes more newsprint than any other country.

Southeast of the Canadian Shield is a narrow region called the St. Lawrence Lowlands. **Lowlands** are low, flat land. The St. Lawrence River is in this region. The Great Lakes and the St. Lawrence River are Canada's busiest **waterways**. Ships carry products through the Great Lakes, to the St. Lawrence River, and then to the Atlantic Ocean. More than half of Canada's people live in the St. Lawrence Lowlands. Toronto and Montreal, Canada's largest cities, are in this region. Canada's capital, Ottawa, is also in this region. Quebec City, the oldest city in Canada, is here, too. It was started by the French in 1608.

The western part of Canada includes the Great Plains and the Rocky Mountains. Wheat is an important

Newspapers being printed on newsprint

The logging industry is found in many parts of Canada. This truck is at a logging site in western Canada.

Wood from Canada's forests is shipped from Vancouver to other countries.

crop on Canada's Great Plains. There are many farms in this region. In some areas people raise cattle. Canada, like the United States, has cowboys to take care of the cattle. Canada's busiest port, Vancouver, is in the west. Vancouver is warmer than Canada's eastern cities. Its water never freezes, so ships use this port all year. From Vancouver, Canada sends goods across the Pacific Ocean.

History, People, and Government

The Indians were the first people to live in Canada. Indian groups are called First Nations in Canada. Then about 5,000 years ago, Inuit people settled in northern Canada. Inuit are also called Eskimos. For thousands of years, Inuit have lived in the Arctic. In the past Inuit hunted and fished for their food. They traveled using dogsleds. Now they travel using snowmobiles and cars. Today Inuit work in mining and oil industries.

An Inuit pilot

Most Inuit live in the eastern part of the Northwest Territories. In 1999 this area will become a new territory called Nunavut. The remaining western part will still be called the Northwest Territories.

In Chapter 1 you read how the French and the British settled Canada. The French were the first people from Europe to settle in Canada. They first

Quebec City

Great Britain's queen, who is also Canada's queen, visits Canada.

settled in the area of Quebec City. Today almost one third of Canada's people are French Canadians. Most of them live in the province of Quebec.

In 1763 the British won control of Canada from the French. Canada became free from Great Britain in 1931. There continues to be a strong friendship between Canada and Great Britain. Canada is part of the **British Commonwealth**. It is a group of nations that were once ruled by Great Britain. Great Britain's queen is also Canada's queen. Almost half of Canada's people are from families that came from Great Britain. Canada also has immigrants from many other nations.

Canada is a democracy. The leader of Canada is the **prime minister**. The people of Canada vote to choose their lawmakers. The lawmakers work in a group called **Parliament**.

Looking at Canada's Future

Many people in Canada care more about their own province than about their country. This is a very big problem in Quebec. Many French Canadians want Quebec to be a separate country. In 1995 people in Quebec voted about separating from Canada. There were enough votes for Quebec to remain part of Canada. But the people of Quebec may vote about separating from Canada again. Canadians must work together to keep Quebec part of their country.

Canada is an important country. It is a rich nation with many resources. Canada works at helping poorer nations. It also sends soldiers to keep peace in other countries. In the years ahead, Canada will continue to be an important country.

Chapter Main Ideas

1. Most Canadians live in southern Canada. Northern Canada is too cold for farms, factories, or cities.
2. The Canadian Shield is a cold region with few people but great natural resources.
3. Canada is a democracy. A prime minister leads the country. Laws are made by Parliament.

Montreal

Imagine being in a French city with 3 million people. The street signs are in French. In many schools the teachers speak only French. But this French city is not in France. It is in the province of Quebec in Canada. The city is Montreal. It is the second largest French city in the world.

Montreal is located on an island. The island is in the St. Lawrence River. It is in southern Quebec. The city was built around a mountain. Montreal is located in the St. Lawrence Lowlands region.

Montreal has the world's largest underground shopping center. The shopping center has more than 200 stores and restaurants. People can go shopping even on the coldest, snowiest days.

Montreal has one of the world's finest subway systems. Trains move quickly under the city's streets. The trains are quiet because they have rubber tires. There is much beautiful artwork in the subway stations. The people of Montreal can enjoy looking at art as they ride to work.

Number your paper from 1 to 5. Write a sentence to answer each question.

1. **Location**   Where is Montreal?

2. **Region**   In which region is Montreal?

3. **Movement**   What do people in Montreal use to travel around the city?

4. **Place**   Why is Montreal a special city?

5. **Human/Environment Interaction**   How has Montreal changed so that people can go shopping in the winter?

◆ Vocabulary

Analogies   An **analogy** compares two pairs of words. The words in the first pair are alike in the same way as the second pair. Number your paper from 1 to 5. Use the words in dark print that best complete the sentences. Write the correct answers on your paper.

waterway   **logging**   **province**   **prime minister**   **permafrost**

1. State is to California as _____ is to Quebec.

2. Sand is to the desert as _____ is to the Arctic.

3. President is to the United States as _____ is to Canada.

4. Road is to car as _____ is to ship.

5. Mining is to metals as _____ is to forests.

◆ Read and Remember

Where Am I?   Number your paper from 1 to 7. Read each sentence. Then look at the words in dark print for the name of the place for each sentence. On your paper write the name of the correct place on the blank after each sentence.

Vancouver   **Canadian Shield**   **St. Lawrence Lowlands**   **Toronto**
Nunavut   **Ottawa**   **Quebec City**

1. "I am in an area where most of the Inuit live."

2. "I am in the capital of Canada."

3. "I am in western Canada in the country's busiest port."

4. "I am in the region that has the cities of Toronto, Montreal, and Quebec."

5. "I am in a cold, rocky region with few people and many natural resources."

6. "I am in the oldest city in Canada."

7. "I am in a city on one of the Great Lakes."

Write the Answer   Number your paper from 1 to 10. On your paper write one or more sentences to answer each question.

1. What is the population of Canada?

2. What is permafrost?

3. What are some of Canada's natural resources?

4. Who is Canada's queen?

5. Which province voted to remain in Canada in 1995?

6. What is the most important crop grown in Canada's Great Plains?

7. Who are the Inuit?

8. Where do most Canadians live?

9. Where is Canada's busiest port city?

10. Who were the first people from Europe to settle in Canada?

◆ Think and Apply

Sequencing   Number your paper from 1 to 5. On your paper write the sentences to show the correct order.

Canada became free from Great Britain in 1931.

The French explored and settled in North America.

People in Quebec voted to remain part of Canada.

Great Britain won control of Canada from the French.

The Inuit first settled in the Arctic.

◆ Journal Writing

Think about the different regions in Canada. Choose the one region where you might want to live. Write a paragraph in your journal that tells why you would want to live in that region.

Maps often use **symbols**, or little pictures, to show information. A **map key** tells what the symbols mean.

A. Number your paper from 1 to 6. Then look at the map key in the map below. On your paper write what each symbol means.

1. ⌒   3. ★   5. •

2. ⊛   4. —   6. ▲

B. Number your paper from 1 to 6. Then study the map key and the map of Canada below. On your paper write the answer to each question.

1. In what province is the capital of Canada?

2. What mountain peak is in the Yukon Territory?

3. Name two cities on the Great Lakes.

4. What is the capital of the province of Alberta?

5. On what river is Montreal located?

6. What province is between Manitoba and Quebec?

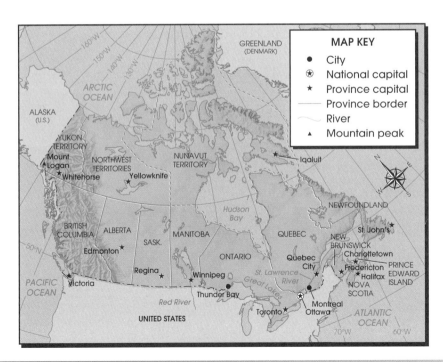

CHAPTER 4

The United States and Canada Work Together

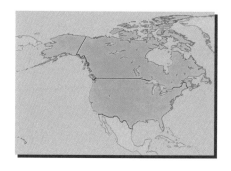

Where Can You Find?
Where can you find a 450-mile waterway that is shared by Canada and the United States?

Think About As You Read

1. How can NAFTA help trade?
2. What are some problems between the United States and Canada?
3. How has the St. Lawrence Seaway helped both countries?

New Words

◆ exports
◆ imports
◆ manufactured goods
◆ NAFTA
◆ tariffs
◆ pollution
◆ acid rain
◆ canals

People and Places

◆ Mexico
◆ Lake Erie
◆ St. Lawrence Seaway

Each year American and Canadian sports teams play many games against each other. Baseball, hockey, and soccer are a few of the sports that build friendship between the two countries.

Trade and NAFTA

Trade is important between the United States and Canada. Each country has more trade with the other than with any other country.

Canada **exports**, or sells to other countries, many products including cars, machines, oil, newsprint, wood, and wheat. Canada **imports**, or buys from other countries, a variety of items. Canada imports fruits, vegetables, and **manufactured goods** from the United States. Manufactured goods are goods that are made in factories.

The United States exports food, machines, cars, and other products to many nations. The United States imports cars, cameras, oil, and other products from

Leaders from Canada, the United States, and Mexico signed the North American Free Trade Agreement, or NAFTA.

President Clinton signed NAFTA in 1993.

other nations. The United States buys many products from Canada.

To improve trade, the United States, Canada, and Mexico signed a trade agreement in 1993. It is called **NAFTA**, or the North American Free Trade Agreement. This agreement helps trade because with NAFTA there are fewer **tariffs**. A tariff is a tax that is paid on imported goods. Tariffs make imported goods more expensive. For example, Canada might put a tariff on shoes that were made in the United States. Then the American shoes would cost more in Canada than shoes that were made in Canada. So Canadians would not buy the American shoes. Tariffs make people want to buy cheaper goods that are made in their own countries. By the year 2005, there will be few tariffs on goods traded between the United States, Canada, and Mexico. Then there will be free trade among the three countries.

NAFTA will make it easier for workers in one country to get jobs in the other countries. It will be easier for each country to start businesses in the other countries. Not everyone likes NAFTA. Some people are afraid that businesses will move from their own country to another country. If businesses move away, many people could lose their jobs. Most people hope that NAFTA will help businesses in all three countries.

Problems Between Neighbors

You learned that Canada is an industrial nation. But many of its factories and businesses are owned by American companies. Many Canadians are not happy that Americans control so many of their businesses.

Many Canadians believe there is too much American culture in their country. Canadians watch American television shows and movies. They listen to American music. But Canadians want their people to pay more attention to their own culture. They also want Americans to learn more about Canada's culture.

Another problem is **pollution**. Pollution means that the air and the water are not clean. Both nations have many cars and factories. These cars and factories send smoke and dirt into the air. This causes air pollution. They also send dirt and chemicals into the water. This causes water pollution. Water pollution hurts fish and plants that live in lakes, rivers, and oceans.

Pollution has hurt the Great Lakes. It did the most harm to Lake Erie. Fish could no longer live in its water. The United States and Canada have worked together to clean the water in the Great Lakes. Both countries passed laws to protect the Great Lakes. The water in Lake Erie is now much cleaner. But still some of the fish in the Great Lakes are not safe to eat.

The two nations are working together to end the problem of **acid rain**. Acid rain forms when pollution in the air becomes part of the rain. Pollution from American factories has caused acid rain in Canada. Canadians are angry because acid rain has killed fish in lakes and rivers. It has killed trees and plants in forests. To stop acid rain, both countries have passed clean air laws. American cars and factories now send much less pollution into the air.

Americans and Canadians Work Together

In 1954 the United States and Canada began building the St. Lawrence Seaway together. The

Pollution in Lake Erie

This leaf shows what happens because of acid rain.

This ship is passing through one of the canals of the St. Lawrence Seaway.

The International Peace Garden is on the border between the United States and Canada.

Seaway is a group of **canals** that go through the rocky part of the St. Lawrence River. The Seaway makes it possible for large ships to sail through the Great Lakes and down the St. Lawrence River. Then ships can sail into the Atlantic Ocean. The trip through the Seaway is about 450 miles long. Ships from many nations use the Seaway to trade with the United States and Canada. Waterpower from the Seaway is used to make electricity. This electricity is used by cities in both nations. The St. Lawrence Seaway has helped both countries.

The United States and Canada share a strong friendship. They also share a long unguarded border. A beautiful peace garden on the border brings people from both countries together.

Chapter Main Ideas
1. The United States and Canada have more trade with each other than with any other country.
2. The United States and Canada are working together to stop air and water pollution and acid rain.
3. The United States and Canada built the St. Lawrence Seaway together. Both countries use electricity from the Seaway.

◆ Vocabulary

Find the Meaning Number your paper from 1 to 5. On your paper write the word or words that best complete each sentence.

1. To **import** goods means to _____ another country.

   sell them to   buy them from   share them with

2. Two kinds of **manufactured goods** are _____.

   apples and corn   water and soil   cars and telephones

3. **Tariffs** make goods from other countries _____.

   cheaper   stronger   more expensive

4. A **canal** is a _____.

   road for cars   track for trains   waterway for ships

5. **Acid rain** forms when rain mixes with _____.

   pollution   snow   mud

◆ Read and Remember

Finish the Paragraph Number your paper from 1 to 5. Use the words in dark print to finish the paragraph below. Write the words you choose next to the correct number on your paper.

Mexico   Great Lakes   imports   sports teams   tariffs

The United States and Canada work together. They have __1__ that play against each other. Both nations have a lot of trade with each other. NAFTA will end most __2__ between the United States, Canada, and __3__. People in these countries want NAFTA to help trade. The United States __4__ many products from Canada. Both nations are working to end pollution in the __5__.

◆ Think and Apply

Find the Main Idea   A **main idea** is an important idea in a chapter. Less important ideas support the main idea. On your paper copy the boxes shown below. Then read the five sentences. Choose the main idea and write it on your paper in the main idea box. Then find three sentences that support the main idea. Write them on your paper in the boxes of the main idea chart. There will be one sentence in the group that you will not use.

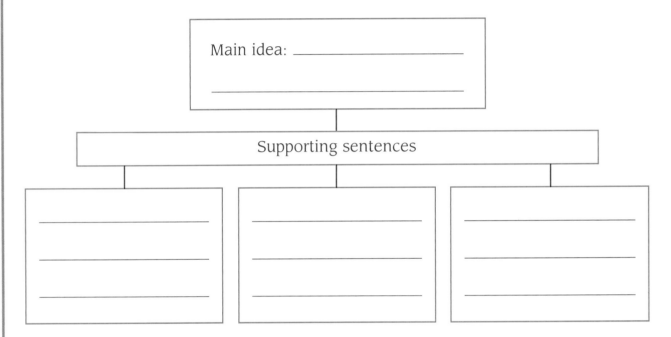

a. Canada exports wheat and oil.

b. The United States and Canada work together in many ways.

c. Both nations have passed laws to stop pollution.

d. The United States and Canada built the St. Lawrence Seaway together.

e. Both nations are working to stop acid rain.

◆ Journal Writing

Imagine you are a newspaper reporter. Write a paragraph that tells about the problems caused by acid rain and pollution. Write at least one way that the United States and Canada are working together to solve the problems.

Mexico City

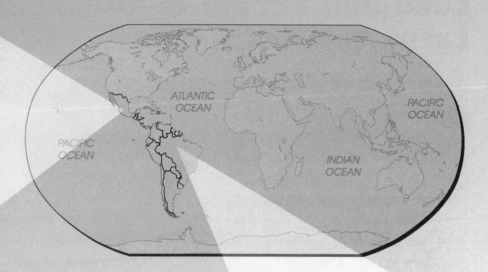

ATLANTIC OCEAN

PACIFIC OCEAN

PACIFIC OCEAN

INDIAN OCEAN

DID YOU KNOW?

- ▲ The Amazon River has enough water to fill a million bathtubs.

- ▲ There are hundreds of volcanoes in Central America.

- ▲ More than 40 ships pass through the Panama Canal every day.

- ▲ More than 6 million people from the United States and Canada visit Mexico each year.

- ▲ Most American baseballs are made in Haiti.

- ▲ Long ago, the Aztec Indians in Mexico played a game that was like basketball.

Amazon River

WRITE A TRAVELOGUE

In this unit you will be visiting many places in Latin America. Keep a travelogue about your trip through this region. Before reading Unit 2, write a paragraph that describes the places in the photographs above. After reading Unit 2, write three or more paragraphs that describe how Latin America is a region.

THEME: REGION

CHAPTER 5

Land and People of Latin America

Where Can You Find?
Where can you find the longest river in Latin America?

Think About As You Read

1. What kinds of landforms and climates are in Latin America?
2. What kinds of people live in Latin America?
3. How do people earn a living in Latin America?

New Words

- Latin
- tropics
- Equator
- Northern Hemisphere
- Southern Hemisphere
- tropical climate
- tropical rain forest
- elevation
- sea level
- mestizos
- developing nations
- subsistence farmers
- plantations

People and Places

- Latin America
- Caribbean Sea
- Middle America
- Central America
- Andes Mountains
- Amazon River
- Portugal
- Catholics
- Venezuela

Which part of the world has the longest mountain chain? Which region has the most Spanish-speaking people? From where in the world do we get most of our coffee and bananas? The answer to all of these questions is Latin America.

Landforms of Latin America

Latin America includes all of the nations south of the United States. This region is called Latin America because most people speak Spanish or Portuguese. Both of these languages developed from **Latin**. Latin is a very old language that few people speak today.

Find Latin America on the map on page 41. Latin America has three parts. The islands in the Caribbean Sea form one part. Middle America is another part. This part includes Mexico and the countries of Central

LATIN AMERICA

Mountains cover large areas of Latin America. What large mountain chain can be found in South America?

America. The third and largest part is the continent of South America.

Mountains cover large areas of Latin America. Some of these mountains are volcanoes. The tall Andes Mountains are in the west of South America. The Andes mountain chain is the longest one in the world. Plains and plateaus cover other parts of Latin America.

There are important rivers in South America. The longest river in South America is the Amazon River. Other large rivers join the Amazon. Many people use these rivers for transportation.

The Tropics

Most of Latin America is in a region called the **tropics**. This is a hot region near the **Equator**. The Equator is an imaginary line that divides the globe in half. The halves are called the **Northern Hemisphere** and the **Southern Hemisphere**. The Equator is the line of latitude that is 0°. The plains near the Equator have a **tropical climate**. A tropical climate is hot all

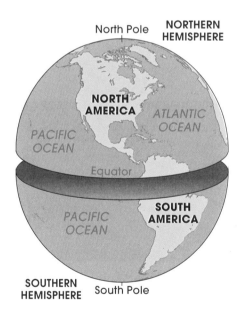

The Equator divides Earth into the Northern Hemisphere and the Southern Hemisphere.

41

These mountains in the tropics are always covered with snow.

An Indian weaver in Central America

the time. It is hot because this region receives more direct sunlight for a longer period of time than other parts of the world. It is also very rainy in the tropics.

Latin America has the world's largest **tropical rain forest**. These thick forests are found near the Equator. They grow where the climate is very hot and very wet. The largest tropical rain forest is in South America. Smaller tropical rain forests can be found in Middle America and on islands in the Caribbean Sea.

Many parts of Latin America have cooler climates even though they are in the tropics. This is because of their **elevation**. Elevation tells you how high the land is above **sea level**. The climate becomes colder as the land's elevation gets higher. Some very tall mountains in the tropics are always covered with snow.

There are two other reasons why some parts of Latin America have cooler climates. First, the southern part of South America is very far from the Equator. As you move away from the Equator, the climate becomes colder. The second reason is that ocean winds give the coastal plains and the islands in the Caribbean Sea milder climates.

History, People, and Culture

Indians were the first people to live in Latin America. Five groups of Indians built great nations in Latin America. They built good roads and large cities.

Five hundred years ago people from Spain conquered most of Latin America. They forced Indians to be their slaves. Millions of Indians died while working for the Spanish. Then the Spanish brought slaves from Africa to work in Latin America.

Spain ruled most of Latin America. But Portugal ruled the huge nation of Brazil. Other countries in Europe also ruled smaller colonies.

In the early 1800s, Latin Americans fought to rule themselves. By 1826 most countries were free.

Who are the people of Latin America today? Some are white people whose families came from Europe. Some people are Indians. Most people are **mestizos**. A mestizo is a person who has European and Indian ancestors. There are also many black people in the Caribbean countries and in Brazil.

The Spanish culture is important in most of Latin America. People in most countries speak Spanish. The Spanish brought the Catholic religion to the region. Today most Latin Americans are Catholics.

Almost three fourths of the people in Latin America live in cities. Some of the world's largest cities are in Latin America. Most of the cities are found on the plateaus and the coastal plains of Latin America. It is hard to live in tall mountains or in tropical rain forests. So these areas of Latin America have small populations.

A Catholic church in Mexico

Earning a Living

Most countries in Latin America are **developing nations**. A developing nation is a nation with a low standard of living and not much industry. In the cities most people work at service jobs. Some people work in factories. Many people in Latin America are farmers.

There are two kinds of farmers in Latin America. There are poor **subsistence farmers** who work on small farms. They try to grow enough food for their families. The second kind of farmer grows cash crops. Coffee, bananas, and sugarcane are important cash crops in Latin America. Cash crops are grown on huge farms called **plantations**. A very small part of the

Subsistence farmers will usually only grow enough food to feed their families. Any extra crops will be sold in a local market.

Latin America has many natural resources. This iron mine is near the Amazon River.

Mine workers in South America

population owns most of the plantations. Millions of poor farmers work on the plantations.

Latin America has many natural resources. Some countries have metals such as silver and copper. In these countries many people work in mines. A few countries like Mexico and Venezuela have oil. But there is little iron and coal in Latin America. These two resources are needed to make cars, trucks, and other machines. This is one reason why there is not much industry in Latin America.

In the next chapters, you will learn more about the land, people, natural resources, and problems of different nations in Latin America. You will find out what people in these nations are doing to solve their problems.

Chapter Main Ideas

1. Many parts of Latin America are covered with mountains and tropical rain forests. The tropical rain forests grow where the climate is very hot and wet.
2. Most people in Latin America speak Spanish and follow the Catholic religion.
3. Most countries in Latin America are developing nations. Many people in Latin America are farmers.

◆ Vocabulary

Match Up Finish the sentences in Group A with words from Group B. On your paper write the letter of each correct answer.

Group A

1. The _____ is half of the globe.

2. A _____ has European and Indian ancestors.

3. _____ grow only enough food for their families.

4. _____ means how high the land is above sea level.

5. The hot region of Earth near the Equator is the _____.

6. _____ have a low standard of living and not much industry.

Group B

A. Subsistence farmers

B. tropics

C. Elevation

D. Northern Hemisphere

E. mestizo

F. Developing nations

◆ Read and Remember

Finish Up Number your paper from 1 to 6. Choose the word or words in dark print that best complete each sentence. Write the word or words on your paper.

Amazon   tropical   bananas   Spanish   Andes   three

1. In most Latin American countries, people speak _____.

2. The world's longest mountain chain is the _____.

3. The longest river in South America is the _____.

4. A _____ climate is hot all the time.

5. Coffee, sugarcane, and _____ are important cash crops in Latin America.

6. Latin America is made up of _____ parts.

Drawing Conclusions   Read the first two sentences below. Then read the third sentence. Notice how it follows from the first two sentences. The third sentence is called a **conclusion**.

There are mountains in Middle America.
There are mountains in South America.
Conclusion: Many parts of Latin America have mountains.

Number your paper from 1 to 4. Read each pair of sentences. Then look in the box for the conclusion you might make. Write the letter of the conclusion on your paper.

1. Tall mountains near the Equator are covered with snow.
Ocean winds cool the islands in the Caribbean Sea.

Conclusion: _____

2. The first people in Latin America were Indians.
Later, people from Europe and Africa came to Latin America.

Conclusion: _____

3. Many people in Latin America have a low standard of living.
Most countries in Latin America do not have much industry.

Conclusion: _____

4. Most people in Latin America speak Spanish.
Most people in Latin America are Catholics.

Conclusion: _____

Conclusions
   A. There are many kinds of people in Latin America.
   B. The Spanish brought their language and religion to Latin America.
   C. Some parts of the tropics have cooler climates.
   D. Many nations in Latin America are developing nations.

◆ **Journal Writing**

Write a paragraph in your journal that tells about landforms, rivers, and tropical rain forests of Latin America.

CHAPTER 6

Mexico: A Nation of Contrasts

Where Can You Find?
Where can you find the largest city in the world?

Think About As You Read

1. What landforms and climates are found in Mexico?
2. How is Mexico becoming an industrial nation?
3. What contrasts can be seen in Mexico?

New Words

- tourists
- colony
- traditional
- tortillas
- rural
- slums
- upper class
- lower class
- middle class
- national debt
- illegal immigrants
- illegal drugs

People and Places

- Rio Grande
- Acapulco
- Central Plateau
- Mexico City
- Aztec
- Father Miguel Hidalgo
- Tenochtitlan

Each winter thousands of Americans leave their cold cities to travel to Mexico's warm beaches. These **tourists** enjoy the many contrasts found in Mexico.

Mexico's Landforms and Climates

Mexico shares its long northern border with the United States. A river called the Rio Grande is part of this border. Mexico has two long coasts with sandy beaches. The beaches of Acapulco are on the Pacific coast. The Gulf of Mexico is to the east.

Tall mountain chains are in the eastern and western parts of the country. Between the two mountain chains is the large Central Plateau. The Central Plateau has a mild climate because of its high elevation. In the north the plateau does not receive much rain. The land is mostly used for cattle ranches. In the south there is more rain and better soil. Many farms are in the south. Part of southern Mexico has a tropical climate. Tropical rain forests are found in this part of Mexico.

Mexico is located between the Pacific Ocean and the Gulf of Mexico. What three countries share borders with Mexico?

MEXICO

Mexico's flag

Most of Mexico's people live on the Central Plateau. Most Mexicans live in cities. Mexico City is the nation's capital. It is the largest city in the world. About twenty million people live in and around Mexico City.

Mexico's History, People, and Culture

Hundreds of years ago, Indians built great nations on the land that is now Mexico. One Indian nation, the Aztec, built schools, temples, and beautiful cities. The Spanish conquered and destroyed the Aztec. Then the Spanish built a **colony** in Mexico. A colony is land that is ruled by another nation.

In the early 1800s, Mexico became the first Spanish colony to fight to be free. Father Miguel Hidalgo started the fight for freedom. By 1821 Mexico was a free country. But Mexicans continued to fight among themselves to control their country.

Father Miguel Hidalgo

In the 1840s, Mexico and the United States fought against each other. The United States wanted Mexican land. Mexico lost the wars. Mexican land to the north of the Rio Grande became part of the United States.

Today Mexico is a democracy. A president leads the nation. Mexico also has a congress.

Who are the people of Mexico today? More than half of the people are mestizos. More than one fourth are Indians. There are also white people whose families once came from Europe.

Spanish and Indian cultures have mixed together to form a Mexican culture. There are many **traditional** villages where people speak Indian languages. Many Indians also speak Spanish. The Spanish language and the Catholic religion are part of Mexico's culture. Mexico is the world's largest Spanish-speaking nation. Mexicans love to watch bullfights. They enjoy spicy foods made with peppers. Mexicans make a special flat bread called **tortillas**, which are made from corn.

Making tortillas

A Developing Nation

Mexico is one of the three nations that signed NAFTA. This trade agreement has helped Mexico sell more products to the United States. It has helped many American companies build new factories in Mexico. The workers in the factories are Mexicans. These factories are helping Mexico become an industrial nation. Mexico is trying to become a developed nation.

Oil is Mexico's most important resource. Mexico also has coal, iron, and silver. Mexico mines more silver than any other country. Mexico earns money by exporting oil and minerals. It sells oil to the United States and to other countries.

Mexico also makes money from tourists. Tourists spend about 4 billion dollars in Mexico each year.

Only a small part of Mexico's land is good for farming. Most of the farmland is used to grow corn. Corn is the main food used by Mexicans. Some cash crops are grown on larger farms. Mexico earns money by exporting cash crops such as sugarcane, wheat, oranges, tomatoes, and bananas.

Tourists in Mexico

In **rural** areas of Mexico, most people are poor farmers. Rural areas are places that are not near cities. Most rural people are subsistence farmers. They grow corn, beans, rice, and squash for their families. Most subsistence farmers have a low standard of living. Many families do not own a car or have a telephone. People may use animals for farm work and for traveling.

Every day many poor people move to Mexico City and to other cities to find better jobs. Millions of poor

Mexico City is the largest city in the world.

Millions of poor people are living in slums in Mexico City.

people live in **slums** that surround these cities. They live in small, poorly made homes.

As in other Latin American countries, most of Mexico's land, businesses, and money are owned by a small group of wealthy people. These rich people are called the **upper class**. The rich people live in large, beautiful homes. Most people in Mexico are poor and belong to the **lower class**. Mexico has a small **middle class** that is not rich or poor. The middle class is slowly growing larger in Mexico.

Solving Problems for the Future

One of Mexico's biggest problems is a huge **national debt**. A country's national debt is the money that it has borrowed and must repay. This problem became very serious in Mexico in 1995. Mexico had borrowed billions of dollars from other countries. But in 1995, Mexico did not have enough money to pay its debts. The United States helped Mexico by lending the country billions of dollars. That money helped Mexico. In 1997 Mexico paid back all the money it had borrowed from the United States.

Each year many poor Mexicans move to the United States. Most have permission to move to the United States. But some Mexicans are **illegal immigrants**. They are people who move to the United States without permission from the American

Many small villages can be found in the hills of southern Mexico.

government. The United States wants Mexico to stop illegal immigrants from entering the United States.

Each year millions of dollars of **illegal drugs** come to the United States from Mexico. It is against the law to sell these drugs in the United States. The United States wants Mexico to work harder to stop illegal drugs from coming into the country.

Mexico is a nation of contrasts. There are contrasts in the landforms and the climate. There are contrasts between modern cities and rural farms. Mexicans are working to improve the standard of living for all of the people of Mexico.

Chapter Main Ideas

1. Most Mexican cities and people are on the Central Plateau.
2. Mexico has many new factories and is trying to become an industrial nation.
3. A large contrast between rich and poor can be seen in Mexico.

Illegal immigrants crossing the border between Mexico and the United States

Mexico City

Imagine living in a city that is more than a mile high. It is surrounded by mountains that are so tall they are always covered with snow. This mile-high city is Mexico City. It is Mexico's capital. It is on the Central Plateau of Mexico. The city has about twenty million people. It is the largest city in the world. It is also the center of Mexico's industry, government, culture, and transportation.

Mexico City first began in the 1300s. At that time the Aztec Indians built a city on the place that is now Mexico City. They built their beautiful city, Tenochtitlan, on an island in a lake. They built bridges to connect their city to the mainland. In 1512 the Spanish conquered the Aztec. They destroyed the Aztec city. The Spanish removed all the water from the lake and built a new city where the Aztec city had been. The new city, Mexico City, became Mexico's capital.

Today Mexico City is very crowded. Poor people from all over Mexico move to Mexico City each day. It also has become a place with terrible air pollution. The city's cars and factories cause pollution. The tall mountains that surround the city prevent the dirty air from moving away from Mexico City. It is no longer healthy to breathe the air in this beautiful capital.

Pollution in Mexico City

Write a sentence to answer each question.

1. **Location**   What is Mexico City's location?

2. **Movement**   How has the movement of people changed Mexico City?

3. **Region**   Why does Mexico City have pollution problems?

4. **Place**   Why is Mexico City a special place?

5. **Human/Environment Interaction**   How did the Spanish change the city of Tenochtitlan?

◆ Vocabulary

Forming Word Groups   Copy the chart shown below on your paper. Read each heading on the chart. Then read each word in the vocabulary list on the left. Form groups of words by writing each vocabulary word under the correct heading on your paper. You should use one answer twice.

Vocabulary List	Understanding Mexico	
upper class	**Types of People**	**Mexico's Problems**
lower class	1. _____	1. _____
illegal drugs		
tourists	2. _____	2. _____
illegal immigrants		
middle class	3. _____	3. _____
city slums		
	4. _____	
	5. _____	

◆ Read and Remember

Find the Answer   Find the sentences that tell about a problem in Mexico. Write the sentences you find on your paper. You should find four sentences.

1. Mexico has a huge national debt.

2. There are millions of poor people.

3. Many tourists go to Acapulco in the winter.

4. Many illegal immigrants go to the United States.

5. Mexico is the largest Spanish-speaking nation.

6. Illegal drugs are sent from Mexico to the United States.

7. Mexico shares its long northern border with the United States.

8. Mexico signed NAFTA with the United States and Canada.

◆ Think and Apply

Cause and Effect   A **cause** is something that makes something else happen. What happens is called the **effect**.

> **Cause:**   Many Mexicans in rural areas need jobs.
> **Effect:**   They move to cities to find work.

Number your paper from 1 to 6. On your paper write sentences by matching each cause on the left with an effect on the right.

Cause

1. Mexico's beaches are warm during the winter, so _____.

2. The southern part of the Central Plateau has good soil, so _____.

3. The Spanish conquered the Indians of Mexico, so _____.

4. Mexico has lots of oil, so _____.

5. The United States won wars against Mexico, so _____.

6. Some Mexicans are rich but most are poor, so _____.

Effect

A. it sells oil to other countries

B. land north of the Rio Grande became part of the United States

C. most farms are in the south

D. many tourists visit Mexico

E. there is a small middle class

F. their cultures have mixed together to form a Mexican culture

◆ Journal Writing

Mexico is a nation of contrasts. It is a developing nation with modern cities and rural farms and villages. Before Mexico can become a developed nation, it must solve its many problems. Write a paragraph in your journal about a problem in Mexico today. Tell how you think the problem could be solved.

Using a Distance Scale

A **distance scale** compares distance on a map with distance in the real world. We use a distance scale to find the distance between two places. Many distance scales show distance in both miles and kilometers.

Look at the map of Mexico below. The distance scale tells us that 1 inch on the map is the same as 600 miles in Mexico. There is about 1 inch between the cities of Chihuahua and Matamoros on the map. So the real distance between the two cities is about 600 miles.

On your paper write the word or numbers that finish each sentence. Use a ruler to measure the distance scale and distances on the map.

1. There is about _____ between Matamoros and Acapulco.

1/10 inch   1/2 inch   1 inch

2. Using the scale, we know that the distance between Matamoros and Acapulco is about _____ miles.

5   125   600

3. There are almost _____ inches between Acapulco and Ciudad Juárez.

1   2   10

4. Using the scale, we know that the distance between Acapulco and Ciudad Juárez is about _____ miles.

100   600   1,050

5. The distance from Ciudad Juárez to Mexico City is about _____ miles.

300   900   2,000

6. The distance from Chihuahua to Mexico City is about _____ miles.

300   750   1,500

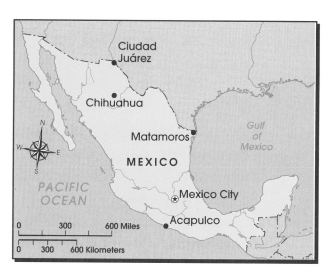

CHAPTER 7

Central America: Seven Countries in the Tropics

Where Can You Find?

Where can you find the only freshwater lake in the world that has sharks?

Think About As You Read

1. What are the important landforms of Central America?
2. Why has Nicaragua had civil wars?
3. How are Costa Rica and Panama different from other Central American nations?

New Words

- highlands
- lava
- earthquakes
- dictators
- civil war
- Communist
- inflation
- isthmus
- democratic

People and Places

- Maya
- Belize
- Guatemala
- Panama
- Nicaragua
- Lake Nicaragua
- Sandinistas
- Cuba
- Soviet Union
- El Salvador
- Honduras
- contras
- Violeta Barrios de Chamorro
- Costa Rica
- Panama Canal

Central America is smaller than the state of Texas. But this region has seven countries. In this chapter, you will learn about life in this region.

The Region of Central America

Central America is located between Mexico and South America. The Pacific Ocean is to the west. The Caribbean Sea is to the east. All seven nations are in the tropics. Find the names of the seven countries on the map on page 57.

Coastal plains are found in the east and the west of Central America. Most cities and plantations are in low mountains called **highlands**. Taller mountains cover a large part of Central America. Some of these mountains are dangerous volcanoes. When a volcano explodes, hot **lava**, or melted rock, pours onto the nearby land. Countries in Central America often have **earthquakes**. During an earthquake the ground shakes and often cracks open. Farms, cities, and

CENTRAL AMERICA

[Map of Central America]

MAP KEY
⭐ Capital city
— Canal
–·– Border between countries

MEXICO
Belmopan
BELIZE
Guatemala City
HONDURAS
GUATEMALA
EL SALVADOR
NICARAGUA
San Salvador
Managua
Tegucigalpa
Lake Nicaragua
Caribbean Sea
PANAMA CANAL
San José
Panama City
COSTA RICA
PANAMA
SOUTH AMERICA
PACIFIC OCEAN

There are seven countries in Central America. Which country is located where Central America and South America meet?

people have been destroyed by volcanoes and earthquakes.

Most people in Central America earn a living through farming. Coffee, bananas, sugarcane, and cotton are the most important cash crops. Most cash crops are grown on large plantations. These plantations are owned by a few rich people. Most people in Central America are very poor. They work on plantations and earn very low salaries. Some people in the region work at mining minerals. Some people work in factories making cotton clothing.

Central America has a long history. More than 2,000 years ago, Indians called the Maya built great cities in this region. In the early 1500s, the Spanish conquered most of Central America. They ruled the region for 300 years. In 1821, six of the seven nations won their freedom from Spain. Belize was a British colony that became free in 1981.

There are different kinds of people in Central America. More than half of the people in Guatemala are Indians. In other countries most people are mestizos or white. There are many African Americans in Panama, Nicaragua, and Belize.

Looking at Nicaragua

Nicaragua is the largest country in Central America. Its landforms and climate are like those of the other

A volcano in Central America

Nicaragua's flag

nations in this region. Nicaragua earns most of its money by selling cotton, coffee, and sugarcane.

Lake Nicaragua is the country's largest lake. It is the only freshwater lake in the world with sharks. One of the islands in Lake Nicaragua has two volcanoes.

For many years **dictators** ruled Nicaragua. A dictator has full power to make laws. These dictators controlled the country's land and money.

Sandinista soldiers

During the 1970s, people called Sandinistas fought a **civil war** against the country's dictator. During a civil war, the people of the same country fight against each other. The Sandinistas won this civil war and took control of the government. The Sandinistas received help from Cuba and the Soviet Union. Those two nations were **Communist** countries and enemies of the United States. In a Communist country, the government controls all businesses. Sandinista soldiers also helped Communists fight in civil wars in El Salvador and Honduras.

The United States did not want Communists to have power in Central America. For a number of years, the United States gave money to people called contras. The contras were against the Sandinistas. The contras and the Sandinistas fought another civil war in Nicaragua. Finally, in 1990 the Sandinistas and the contras agreed to allow free elections in Nicaragua. The civil war ended.

Violeta Barrios de Chamorro

The people elected Violeta Barrios de Chamorro to be president. People voted for Chamorro because she was against the Sandinistas. But as president she allowed some Sandinistas to have power in the government.

Nicaragua has had many problems since its civil war ended. The long war destroyed many parts of the country. Many people do not have jobs. Most people earn very little money. **Inflation** is another big problem. Inflation means that food and goods become more and more expensive to buy. Inflation makes people need more money to buy almost everything.

These bananas were grown on a plantation in Costa Rica.

Costa Rica's flag

Peaceful Costa Rica

Costa Rica is south of Nicaragua. It is a peaceful country. It has never had a civil war. Costa Rica does not even have an army. Costa Rica is also different because its government is a strong democracy.

Costa Rica has the highest standard of living in Central America. Most people work on small farms, but they have a comfortable life. The country exports cash crops such as coffee and bananas. It also earns money from tourists. Costa Rica has more factories than other nations in the region. Factory workers make food products, clothing, machines, medicines, and other goods. The people of Costa Rica are proud of their peaceful democracy.

Panama and the Panama Canal

Panama is south of Costa Rica. It joins Central America to South America. Panama is an **isthmus** between the Atlantic and Pacific oceans. An isthmus is a narrow piece of land that is between two large bodies of water.

During the early 1900s, the United States built a canal across Panama. To build the Panama Canal, thousands of workers cut through mountains and rain forests. The Panama Canal allows ships to sail from the Atlantic Ocean through Panama to the Pacific

San José is Costa Rica's capital.

The Panama Canal saves weeks of travel time. By sailing through the Canal, ships do not have to go around South America.

Flag of Panama

The Panama Canal

Ocean. Panama earns a lot of money from ships that use the Canal. The United States has controlled the Canal since it was built. But at the end of 1999, Panama will have control of the Canal.

Working for a Better Future

To help the people of this region, more people must own land. Once people own land, their government could help them get seeds and farm machines so they can be better farmers. The region also needs more industry so that countries in the region can sell more products other than cash crops.

Civil wars have been a big problem in many Central American nations. The nations of Central America need strong **democratic** governments. They also need peace. Peace will help all people of the region to have a better life.

Chapter Main Ideas

1. Large areas of Central America are covered with mountains. Most people live in highlands.
2. The Sandinistas lost the civil war in Nicaragua. The people elected a new president in 1990.
3. Costa Rica is a peaceful democracy. Costa Rica has the highest standard of living in Central America.

◆ Vocabulary

Finish the Paragraph Use the words in dark print to finish the paragraph below. On your paper write the words you choose.

dictator	**earthquakes**	**isthmus**
civil war	**highlands**	**inflation**

In Central America most people and cities are in the low mountains called ___1___. This region sometimes has ___2___ in which the ground shakes and cracks open. During a ___3___ the people of the same country fight against each other. Before 1990 a ___4___ ruled Nicaragua and had full power to make laws. Food and goods in Nicaragua continue to become more expensive. This problem is called ___5___. Panama is an ___6___ because it is a narrow piece of land between two large bodies of water.

◆ Read and Remember

Write the Answer On your paper write one or more sentences to answer each question.

1. How do most people in Central America earn money?

2. What did the Maya do in Central America long ago?

3. How did the civil war end in Nicaragua in 1990?

4. What are two ways that Costa Rica is different from Nicaragua?

5. Look at the map of Central America on page 57. What are the nations of Central America?

6. How has the Panama Canal helped Panama?

7. What will happen to the Panama Canal by the end of 1999?

8. Why is inflation a problem for the people of Nicaragua?

Fact or Opinion   A **fact** is a true statement. An **opinion** is a statement that tells what a person thinks.

>**Fact**   Costa Rica does not have an army.
>**Opinion**   Costa Rica should have an army.

Number your paper from 1 to 10. Write **F** on your paper for each fact. Write **O** for each opinion. You should find five sentences that are opinions.

1. Coffee, bananas, sugarcane, and cotton are cash crops in Central America.

2. Cash crops are grown on large plantations.

3. Costa Rica has the highest standard of living in Central America.

4. There are too many farmers in Central America.

5. People should not live near the Panama Canal.

6. Costa Rica is a democracy.

7. Many people in Nicaragua do not have jobs.

8. The Maya built better cities than the Aztec.

9. The Panama Canal makes it easier for ships to sail from the Atlantic Ocean to the Pacific Ocean.

10. The United States should keep control of the Panama Canal after 1999.

11. Volcanoes and earthquakes can cause a lot of damage.

12. Inflation should not be allowed to happen.

◆ Journal Writing

Nicaragua and Costa Rica are two countries in Central America. Like any two things, these countries have some things that are the same and some things that are different. Write a paragraph in your journal that compares Nicaragua and Costa Rica. Tell how the two countries are alike. Tell how they are different.

The map below is a **grid map** for part of Panama City, the capital of Panama. **Grids** are often drawn on maps to help find places on the map. Each place on the map is in a grid square. On this map the Museum is located in square B-2.

Number your paper from 1 to 5. Then look at the map below. Answer each question.

1. What place is in square E-2?

2. What place is in B-3?

3. What road is in C-1?

4. In which grid square is the Supreme Court of Justice located?

5. What road is in B-4?

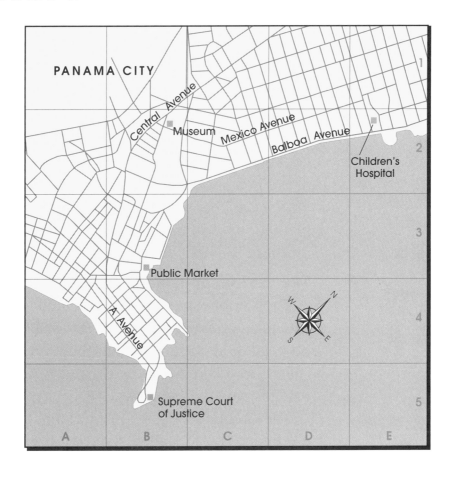

CHAPTER 8

The Caribbean: Hundreds of Beautiful Islands

Where Can You Find?
Where can you find the oldest city started in the Western Hemisphere by Europeans?

Think About As You Read

1. How do people in the Caribbean Islands earn a living?
2. In what ways have cultures mixed together in the Caribbean Islands?
3. In what ways are Cuba, Haiti, and the Dominican Republic alike? How are they different?

New Words

- ◆ tropical storms
- ◆ hurricanes
- ◆ tourism
- ◆ bauxite
- ◆ aluminum
- ◆ calypso
- ◆ independent
- ◆ Western Hemisphere
- ◆ Eastern Hemisphere

People and Places

- ◆ Caribbean Islands
- ◆ Greater Antilles
- ◆ Lesser Antilles
- ◆ Bahamas
- ◆ Jamaica
- ◆ Christopher Columbus
- ◆ Hispaniola
- ◆ Haiti
- ◆ Dominican Republic
- ◆ Fidel Castro
- ◆ Toussaint L'Ouverture
- ◆ Jean-Bertrand Aristide
- ◆ Santo Domingo

Imagine you are on an island when it begins to rain. You run under a tree with giant leaves. Soon the rain stops and the sun comes out again. There is a cool breeze. Workers return to the fields. More tourists arrive in a ship. You have imagined some parts of life on the Caribbean Islands. Beautiful beaches, large plantations, and many different cultures can be found on these islands.

The Caribbean Islands

The Caribbean Islands are located between the Atlantic Ocean and the Caribbean Sea. Look at the map on page 65. You can see that there are three groups of islands in this region—the Greater Antilles, the Lesser Antilles, and the Bahamas.

The Greater Antilles has four large islands. There are hundreds of smaller islands in the Lesser Antilles. Mountains cover many of these islands. Some of these

THE CARIBBEAN ISLANDS

THE CARIBBEAN ISLANDS

Nassau

Havana

BAHAMAS

CUBA

ATLANTIC OCEAN

GREATER

HAITI DOMINICAN REPUBLIC

San Juan

JAMAICA Kingston

ANTILLES

PUERTO RICO (U.S.)

ANTIGUA AND BARBUDA

DOMINICA

Caribbean Sea

ANTILLES

ST. LUCIA

BARBADOS

MAP KEY
- City
⊛ Capital city
-·- Border between countries

LESSER

GRENADA

Port-of-Spain TRINIDAD

There are three groups of islands in the Caribbean region. The largest islands in the Caribbean are in the Greater Antilles. What is the largest island in this region?

mountains are volcanoes. The Bahamas and some of the Lesser Antilles islands are mostly flat.

Most of the Caribbean Islands are in the tropics. Cool winds from the sea keep the islands at about 80°F all year. **Tropical storms** are a problem between July and October. Tropical storms bring heavy rains and strong winds. The worst tropical storms are **hurricanes**. Hurricane winds can tear trees out of the ground and blow down houses. Rainfall during a hurricane often causes floods.

Every year millions of tourists visit the Caribbean Islands. Many people work in service jobs in the **tourism** industry. They work in hotels, restaurants, and stores that tourists visit.

Most people in the Caribbean Islands work in farming. The Caribbean Islands earn most of their money from cash crops. Sugarcane is the most important cash crop. Bananas, oranges, rice, and pineapples are also exported. Most Caribbean Island countries have to import food for their people.

A hurricane passed through this Caribbean island.

There is little industry in the Caribbean Islands. But some countries have factories where food products are made. In some countries, people work in mines. Many people in Jamaica work in **bauxite** mines. A metal called **aluminum** is made from bauxite.

The Culture of the Caribbean

Christopher Columbus explored the Caribbean for Spain. Later, Spain started colonies on several islands including Cuba, Puerto Rico, and Hispaniola. Other European countries also started colonies. They brought African slaves to work on plantations. Today most people in the Caribbean have ancestors who were brought to the islands from Africa.

A calypso band

Different cultures from Europe and Africa mixed together in the Caribbean Islands. For example, **calypso** music is a mix of African, Spanish, and American music. Today Spanish, French, English, and other languages are spoken on different islands. Some people speak Caribbean languages that use words from African and European languages.

Many Caribbean Islands are **independent** countries. Some islands do not rule themselves. For example, Puerto Rico is not independent. It belongs to the United States. Let's look at three Caribbean countries: Cuba, Haiti, and the Dominican Republic.

Cuba: A Communist Country

Cuba is close to the United States. It is only 90 miles from Florida. But the United States and Cuba are not

Many different languages can be heard in this marketplace in the Dominican Republic.

Sugarcane is cut by hand on this plantation in Cuba. Sugar is Cuba's most important export.

Flag of Cuba

friendly neighbors. They are enemies because Cuba is a Communist country. Cuba is the only Communist country in the **Western Hemisphere**. North America and South America are in the Western Hemisphere. Asia, Africa, Europe, and Australia are continents in the **Eastern Hemisphere**.

Before 1959 many businesses and sugarcane plantations in Cuba were owned by Americans. Many times the United States sent soldiers to Cuba to protect American businesses. Since 1959 a dictator named Fidel Castro has ruled Cuba. He started a Communist government in Cuba. Castro took control of all the American businesses and plantations in Cuba. They are now owned by Cuba's government.

Many Cubans were unhappy with Castro. About one million Cubans have moved to the United States. Others have moved to Mexico and Spain.

Today Cuba has many problems. Cuba gets little money from tourism. Cubans do not have enough food or money. But Castro has improved schools, roads, and health care.

Fidel Castro

Haiti and the Dominican Republic

Haiti is in the western part of Hispaniola. Haiti is covered with mountains. There is not much farmland. Most people in Haiti are poor farmers. Their farms are very small, so there is not enough food.

Haiti and the Dominican Republic are located on the island of Hispaniola.

Haiti began as a French colony. The French brought many African slaves to work on their plantations. In 1791 an African slave, Toussaint L'Ouverture, began a fight for freedom. Haiti became independent in 1804.

Haiti has been ruled by many dictators. Since 1990 Haiti has tried to become a democracy. The country held free elections. The people elected Jean-Bertrand Aristide to be president. Haiti's army forced the new president to leave. But in 1994 the United States sent soldiers to Haiti to help Aristide rule his country. In 1995 the people of Haiti voted for a new president.

The Dominican Republic is on the eastern part of Hispaniola. Its capital, Santo Domingo, is the oldest city started in the Western Hemisphere by Europeans. Spanish is the language of the Dominican Republic.

The Dominican Republic is larger than Haiti. It also has more farmland. Sugarcane is the main cash crop. This country is more developed than Haiti. There are more factories. The Dominican Republic also earns money from tourism.

The Future of the Caribbean Islands

All countries in the Caribbean Islands are developing nations. Too many people are poor farmers. These countries need to grow more food to feed their people. Most Caribbean countries depend too much on sugarcane to earn money. Sometimes the price of sugar drops. Then these countries cannot earn enough money. Millions of people do not have jobs. New industries will give more people jobs. The Caribbean is a beautiful region. But the people of this region need a higher standard of living.

Jean-Bertrand Aristide

Chapter Main Ideas

1. Different cultures from Europe and Africa mixed together in the Caribbean Islands. The people speak many languages and have different cultures.
2. Sugarcane is the most important cash crop on most Caribbean Islands. There is not much industry.
3. Cuba, Haiti, and the Dominican Republic are developing nations.

◆ Vocabulary

Analogies   Use the words in dark print that best complete the sentences. Write the correct answers on your paper.

aluminum   Western Hemisphere   hurricanes   calypso

1. Sugarcane is to cash crop as _____ is to music.

2. Europe is to Eastern Hemisphere as North America is to _____.

3. Oranges are to orange juice as bauxite is to _____.

4. The Arctic is to snowstorms as the Caribbean is to _____.

◆ Read and Remember

Where Am I?   Read each sentence. Then look at the words in dark print for the name of the place for each sentence. Write the name of the correct place on your paper.

Dominican Republic   Haiti   Puerto Rico   Hispaniola
Santo Domingo   Cuba   Jamaica   Bahamas

1. "I am in a country with many bauxite mines."

2. "I am in a place that belongs to the United States."

3. "I am in a Communist country, and it is 90 miles from Florida."

4. "I am in a poor country that held free elections in 1990."

5. "I am in the oldest European city in the Western Hemisphere."

6. "I am in a Spanish-speaking country that is in the eastern part of Hispaniola."

7. "I am on the Caribbean island that contains two independent countries."

8. "I am in the northernmost group of Caribbean islands."

Finish Up Choose the word or words in dark print that best complete each sentence. Write the word or words on your paper.

tourism **tropical storms** **Greater Antilles** **developing**
farming **volcanoes** **sugarcane** **bauxite**

1. There are four large islands in the _____.

2. Some of the mountains in the Caribbean Islands are _____.

3. Between July and October there are many _____ in the Caribbean.

4. Many people work in service jobs in the _____ industry.

5. Most people in the Caribbean Islands work in _____.

6. The most important cash crop in the Caribbean Islands is _____.

7. Many people in Jamaica work in _____ mines.

8. All of the countries in the Caribbean Islands are _____ nations.

◆ Think and Apply

Sequencing Number your paper from 1 to 5. Write the sentences on your paper to show the correct order.

In 1990 the people of Haiti elected Jean-Bertrand Aristide to be president.

Fidel Castro became president of Cuba.

Toussaint L'Ouverture led the fight for freedom in Haiti.

Spain and other European countries started colonies in the Caribbean.

The United States Army helped Jean-Bertrand Aristide return to Haiti to become president again.

◆ Journal Writing

You have read about the island nations in the Caribbean area of Latin America. Write a paragraph in your journal that compares Haiti and Cuba. Tell how they are alike and how they are different.

Understanding Lines of Latitude

You have read that the Equator is an imaginary line around the center of Earth. It runs east and west. There are many more lines that run east and west around Earth. These lines are called **lines of latitude**. Lines of latitude help us find places on maps and globes. They form part of a grid.

Each line of latitude is named with a number of **degrees**. The Equator is 0°. All other lines of latitude are north or south of the Equator. There are 90 lines of latitude in the Northern Hemisphere. The North Pole is 90° North, or 90°N. There are 90 lines of latitude in the Southern Hemisphere. The South Pole is 90° South, or 90°S.

Look at the map of Middle America and the Caribbean below. Then on your paper write the correct answer to finish each sentence.

1. The latitude of Mexico City is about
———.

15°S   20°N   80°N

2. The latitude of San José, Costa Rica, is ———.

10°N   10°S   40°S

3. The latitude of Jamaica is about ———.

30°N   18°N   8°S

4. A city with a latitude of 20°N is ———.

Santiago   Ciudad Juárez

Panama City

CHAPTER 9

Brazil: Home of the Amazon Rain Forest

Where Can You Find?

Where can you find a river that is almost 100 miles wide?

Think About As You Read

1. How is Brazil different from other Latin American countries?
2. What is happening to the Amazon rain forest?
3. What kinds of products does Brazil export?

New Words

- mouth
- mulattos
- Carnaval
- interior
- basin
- deforestation
- oxygen
- poverty
- homeless

People and Places

- São Paulo
- Rio de Janeiro
- Brasília
- Amazon rain forest
- Amazon River basin
- Chico Mendes

Where can you find the longest river in the Western Hemisphere? Where can you find the world's largest rain forest? The answer to both questions is Brazil.

Brazil's Landforms

Brazil is the largest country in Latin America. Most of Brazil has a tropical climate. The world's largest tropical rain forest covers about half of Brazil. The Amazon River passes through this forest.

The Amazon is the world's second-longest river. It starts in the Andes Mountains in the western countries of South America. In Brazil the Amazon runs for more than 2,000 miles. Then it flows into the Atlantic Ocean. Ocean ships travel through Brazil on the Amazon. Many shorter rivers flow into the Amazon. The Amazon has a lot of water. In some places the river is several miles wide. The **mouth** of the Amazon is almost 100 miles wide. The mouth is where the river flows into the ocean.

BRAZIL

The Amazon River in Brazil is one of the world's longest rivers. What is the location of the mouth of the Amazon River?

Brazil's flag

Brazil has a wide coastal plain near the Atlantic Ocean. A large plateau covers the southern part of the country. This plateau has a milder climate than the northern part of Brazil. Most cities and farms are on the southern plateau.

Brazil's People, Cities, and Resources

Brazil has more people than any other country in Latin America. People from Portugal settled in Brazil in the 1500s. They brought the Catholic religion and the Portuguese language to Brazil. Today more than half of Brazil's people are white. There are also many black people and **mulattos**. A mulatto is a person with black and white ancestors. The Indian population in Brazil is small.

Most people in Brazil live within 200 miles of the Atlantic Ocean. Most of Brazil's people live in cities near the ocean. São Paulo is Brazil's largest city. It is one of the largest cities in the world.

Brazil is famous for its **Carnaval** holiday. During Carnaval people wear costumes and march and dance in parades. The most exciting Carnaval is in the city of Rio de Janeiro. Many tourists visit Rio for Carnaval.

Dancers in costume for Carnaval in Rio de Janeiro

From above, the rain forest looks like a thick, green blanket.

Brazil's government wants people to develop the western part of Brazil. This western region is called the **interior**. To develop the interior, Brazil built a new capital city there. You read about Brazil's capital, Brasília, in the introduction to this book.

Brazil has many natural resources. It has bauxite, gold, tin, iron, and other metals. There is waterpower for making electricity. Brazil also has some oil. But many natural resources are in the interior in the tropical rain forest. So they are hard to reach.

Brazil's Amazon Rain Forest

The Amazon rain forest covers about half of Brazil. It also covers parts of other South American countries. It is the world's largest rain forest. This forest is in the Amazon River **basin**. A basin is the area that drains into a river. Low plains surround the Amazon River.

What is it like to be in the rain forest? Thousands of different kinds of plants and animals live in the rain forest. Tall trees are everywhere. Some trees are hundreds of feet tall. Water drips off of the trees all the time because it rains so often. The ground is often wet and muddy. The climate in the tropics is hot. But the air in the rain forest feels cooler because the tall trees block the hot sun. Thick plants cover the ground where the sunlight shines through the tall trees. Plants need this sunlight to grow.

The Amazon rain forest is the largest one in the world. Many different types of plants and animals are found in this rain forest.

The rain forest has many resources. Nuts come from its trees. Rubber trees in the rain forest provide rubber for tires. Gold, tin, and other metals are in the earth. Medicines are made from trees and plants in the rain forest. No other part of the world has the trees and plants needed to make these medicines. These medicines help people who are very sick.

Deforestation

Brazil's rain forest is being destroyed. Destroying forests is called **deforestation**. Deforestation harms Earth's land, air, and water. People are cutting down trees to build roads through the forest. The new roads make it easier to travel through the region. People cut down the rain forest trees to mine metals and to have new farmland. After cutting the trees, they burn them in order to clear the land. There are now thousands of fires each day as more land is cleared. These fires cause terrible air pollution.

Trees in the rain forest are cut down to clear the land.

People have started farms and cattle ranches on cleared land. But the soil in the rain forest is not good for farming. It is also not good for raising cattle. After a few years, farmers cannot raise good crops on rain forest land. Then they move to another part of the rain forest. There they cut and burn more trees. They try again to farm the land. Each time this happens, more of the rain forest is destroyed. Most animals, plants, and trees of the rain forest can never be replaced. Small groups of Indians live in the rain forest. They use the rain forest, but they do not destroy it.

The rain forest is important to people in all parts of the world. Its trees send a gas called **oxygen** into the air. This oxygen makes the air cleaner and healthier to breathe. Many nations want Brazil to save its rain forest. People everywhere depend on oxygen and medicines from the rain forest.

Burning the rain forest

Working for a Better Future

Brazil is working to become a developed nation. Cars, machines, and shoes are a few of Brazil's factory products. Brazil exports factory goods and cash crops. Much of the world's coffee and sugar come from Brazil.

Homeless child in Rio de Janeiro

Poverty is a big problem in Brazil. Most plantation workers are poor. Slums surround Brazil's cities. Many children are **homeless**, or without homes. Brazil's government must find ways to help millions of poor people to have a higher standard of living.

The world is watching Brazil to see if it will save the Amazon rain forest. Brazil says it needs the forest's resources to help people earn more money. It is possible to get resources from the forest and not destroy it. By saving the rain forest, Brazil can use its resources to build a better future.

Chapter Main Ideas

1. Brazil is the largest country in Latin America.
2. People are destroying the Amazon rain forest in order to use its land and resources.
3. Brazil exports many factory goods and cash crops.

BIOGRAPHY

Chico Mendes (1944-1988)

Chico Mendes grew up in a poor family in Brazil. He became a leader among the rubber tappers in the Amazon rain forest. Rubber tappers remove rubber from rubber trees without harming the trees. Mendes became angry when people began chopping down rain forest trees. Rubber tappers could not work if the rubber trees were gone. So Mendes began his fight to save the rain forest.

Ranchers who wanted land in the Amazon rain forest hated Mendes. They said they would give him money if he stopped trying to save the trees. But Mendes continued his fight. He gave many speeches. People around the world learned why Brazil must save its rain forest.

On December 22, 1988, Mendes was killed in his own home by some ranchers. Other people are now doing the work Mendes started. They continue to fight for the Amazon rain forest.

Journal Writing
Write a paragraph in your journal that tells why Chico Mendes is a hero to many people.

◆ Vocabulary

Finish Up   Choose the word in dark print that best completes each sentence. Write the word on your paper.

deforestation   oxygen   homeless   interior   basin

1. The western area of Brazil is the _____.

2. The area where water drains into a river is a river _____.

3. When a forest is destroyed, a region has _____.

4. We must breathe a gas in the air called _____.

5. People who do not have homes are _____.

◆ Read and Remember

Complete the Geography Organizer   Copy the geography organizer shown below on your paper. Complete it with information about Brazil.

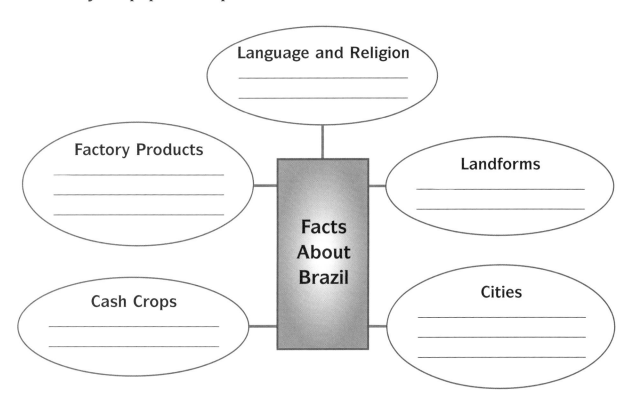

Matching   Each item in Group B tells about an item in Group A. Write the letter of each item in Group B next to the correct number on your paper.

Group A

1. Carnaval

2. Amazon

3. medicines

4. farms and cattle ranches

5. Brasília

Group B

A. This is the world's second longest river.

B. During this holiday, people wear costumes and dance in parades.

C. This capital city was built to help develop Brazil's interior.

D. The rain forest is being destroyed to make room for these.

E. These are made from trees and plants in the rain forest.

◆ Think and Apply

Finding Relevant Information   Information that is **relevant** is information that is important for what you want to say or write. Imagine you are telling your friend why it is important to save the Amazon rain forest. Read each sentence below. Decide which sentences are relevant to what you will say.

On your paper write the relevant sentences you find. You should find three relevant sentences.

1. Trees in the rain forest send oxygen into the air.

2. Medicines are made from plants and trees in the rain forest.

3. Many minerals can be mined from below the rain forest's ground.

4. Rain forests are located in the tropics.

5. Indians need the rain forest in order to live.

◆ Journal Writing

Do you think the rain forest should be saved? Write a paragraph in your journal that explains your opinion.

Understanding Lines of Longitude

You learned that lines of latitude run east and west around Earth. There are also lines that run north and south around Earth. These lines are called **lines of longitude**. All lines of longitude meet at the North Pole and the South Pole. Lines of longitude are named with their number of degrees, just like lines of latitude. The **Prime Meridian** is a line of longitude that runs through Greenwich, England. The Prime Meridian is 0 degrees, or 0°. All other lines of longitude are east or west of the Prime Meridian. There are 180 lines of longitude to the west and 180 lines to the east.

Look at the map of Brazil below. On your paper write the answer to each question.

1. What is the longitude for São Paulo?

    120°E   47°W   40°E

2. What is the longitude for Brasília?

    48°W   148°E   10°E

3. What is the longitude of Manaus?

    40°W   50°W   60°W

4. What city has the longitude of 35°W?

    Manaus   Recife   Rio de Janeiro

5. How many large cities in Brazil are to the west of 60°W?

    0   5   15

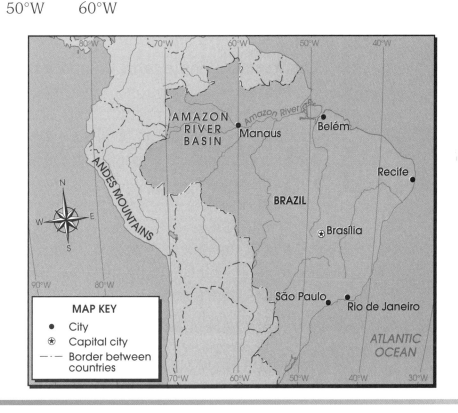

MAP KEY
- City
- Capital city
- Border between countries

CHAPTER 10

The Countries of the Andes

Where Can You Find?

Where can you find the capital city with the highest elevation in the world?

Think About As You Read

1. What is the main landform of the Andean countries?
2. How does elevation change the climate in the Andes?
3. What problems are found in this region?

New Words

- empire
- Quechua
- coca
- cocaine
- drug traffickers
- terrorism
- terrorists
- ambassador
- marijuana

People and Places

- Andean countries
- La Paz
- Bolivia
- Peru
- Inca
- Lima
- Japan
- Colombia
- Alberto Fujimori
- Chile

Where can you find the world's longest mountain chain? Where can you find the world's longest country? The answer is in the countries of the Andes Mountains.

Landforms, People, and Climates of the Andes

There are five countries in the Andes Mountains. They are called the Andean countries. Find the five countries on the map on page 82. Look at how the Andes mountain chain is in all of these countries. These mountains are found along the west coast of South America. This mountain chain is more than 4,000 miles long.

Most of the people who live in the Andes Mountains are Indians, and some are mestizos. Most people are poor subsistence farmers.

The capital city with the world's highest elevation is in the Andes. It is La Paz, Bolivia. This capital is more than 12,000 feet high. The climate in La Paz is cool because of the high elevation. Most Andean

FOUR CLIMATE REGIONS IN THE ANDES

Frozen Land, Tierra Helada
(above 15,000 feet)

Cold Land, Tierra Fria
potatoes
(7,000–15,000 feet)

Mild Land, Tierra Templada
coffee, corn, wheat
(3,000–7,000 feet)

Hot Land, Tierra Caliente
sugarcane, coconuts, bananas
(0–3,000 feet)

Sea Level (0 feet)

The English and Spanish names for each region describe the region's climate. What crops grow in the Hot Land climate?

countries are in the tropics. Places in the tropics have hot, tropical climates. But in some parts of the Andean nations, the climate is not hot. The climate gets cooler as the elevation gets higher. The chart above shows four climate regions in the Andean countries.

Peru: The Largest Andean Country

Peru is the third largest country in South America. It is also the largest country in the Andes.

For thousands of years, Indians have lived in Peru. About the year 1200, the Inca Indians built a great **empire** in the Andes. An empire is an area where several nations are ruled by one nation. The Inca built cities and excellent roads. Their empire lasted more than 300 years.

The Spanish conquered the Inca in the 1500s. They built a Spanish colony in Peru. The Spanish built Peru's capital, Lima. It is on the coast. From Lima the Spanish sent Peru's minerals on ships to Spain.

In the 1820s Peru became free from Spain. Today Peru has two official languages. Spanish is one language. **Quechua**, an Indian language, is the other.

Today almost half of Peru's people are Indians. The Inca were their ancestors. Peru has the largest Indian population in South America. Many people in Peru are mestizos. Peru also has a small white population. There are some immigrants from Japan and other countries in Asia, too.

This is one of the places in Peru where the Inca once lived.

The Andes mountain chain is located along the length of western South America. In which countries are the Andes located?

Map: Countries of the Andes

80°W · 70°W · 50°W · 40°W · 30°W

10°N

Bogotá ⊛
COLOMBIA

Equator

Quito ⊛
ECUADOR

0°

PERU

BRAZIL

ANDES MTS.

10°S

Lima ⊛
Lake Titicaca
La Paz ⊛
BOLIVIA

ATACAMA DESERT

20°S

PACIFIC
OCEAN

ATLANTIC
OCEAN

Santiago ⊛

30°S

CHILE

40°S

MAP KEY
⊛ Capital city
—·— Border between countries

100°W · 90°W · 80°W · 60°W

Farmer selling coca leaves

The small white population is Peru's upper class. The upper class owns most of Peru's land, businesses, and plantations. Most mestizos are in the lower class. They work on farms and in factories on the coastal plain. The Indians are the poorest people in Peru. Most are subsistence farmers. They raise potatoes, wheat, and corn in the mountains.

Many subsistence farmers grow a plant called **coca**. Coca is used to make an illegal drug called **cocaine**. In Peru and Colombia, people called **drug traffickers** pay the farmers to grow coca. Farmers earn more money from coca than from any other crop. Drug traffickers earn a lot of money by selling cocaine in the United States.

Peru has many natural resources. It has silver, tin, and other metals. There is also iron and oil. Many

Terrorists took over the home of Japan's ambassador to Peru.

minerals are found in the Andes. But they are hard to reach because they are in the mountains. Peru does not have enough good roads. It is hard to build roads through the Andes. The mountains make it difficult to travel from one place to another.

Terrorism is a very big problem in Peru. Terrorism is the use of dangerous acts against the people of a country. **Terrorists** try to destroy their enemies with these dangerous acts. In the 1980s and 1990s terrorists have tried to win control of Peru's government. They have killed about 30,000 people. In December 1996, terrorists forced hundreds of guests at the home of Japan's **ambassador** to be their prisoners. After four months, Peru's army was able to free the prisoners.

In 1990 Alberto Fujimori was elected president of Peru. He has tried to stop terrorism in Peru. He has also improved the country's economy. Peru had a terrible inflation problem. Peru's economy has been growing stronger. Many more people now work in factories and at jobs in cities. Inflation is not a problem in Peru today.

Alberto Fujimori

Chile: A Long, Narrow Country

Chile is south of Peru. It is more than 2,000 miles long. But it is only about 100 miles wide. The climate is colder in southern Chile than in northern Chile.

Most of Chile's cities and people are in a region called the central valley. The soil and the climate are good for farming. Chile is in the Southern Hemisphere.

Coffee beans

When it is winter in the Northern Hemisphere, it is summer in the Southern Hemisphere. Chile's farmers grow fruits and vegetables during the summer. People in North America enjoy this summer fruit from Chile during the North's winter.

Colombia: Coffee Exports and More

Colombia is north of Peru. Colombia is the only country in South America with coasts on both the Pacific Ocean and the Caribbean Sea.

For many years most of Colombia's money came from exporting coffee. Now Colombia also has an oil industry. One day Colombia might earn more from oil than from coffee. Colombia also exports fresh flowers, bananas, and minerals.

Today people in Colombia earn more money from illegal drugs than from all other products. Drug traffickers are producing **marijuana** and cocaine. They sell these illegal drugs in the United States. Many crimes in Colombia and other countries are caused by these drug traffickers. The United States wants Colombia, Peru, and other countries to stop the work of drug traffickers.

The Future of the Andean Countries

The Andean countries have many poor people. Upper class people own most of the region's wealth. Governments must help the middle class grow larger.

The Indians of the Andes want the same rights and power as white people. Bolivia became the first country to elect a president who is an Indian. But most Indians still have fewer rights and less power.

The Andean countries need peace. They need peace from terrorists. They need peace from drug traffickers. Peace will help people work for a better future.

Indians in Peru

Chapter Main Ideas

1. The Andes Mountains are along the west coast of South America.
2. The Andean countries have four different climates because of elevation.
3. The Andean countries have had problems with inflation, terrorism, and drug traffickers.

◆ Vocabulary

Match Up   Finish the sentences in Group A with words from Group B. Write the letter of each correct answer on your paper.

Group A

1. Peru has two official languages, Spanish and _____.

2. Many people have been killed by _____ who have used dangerous acts to try to win control of Peru.

3. The Inca built an _____ that lasted for 300 years.

4. In Colombia, people called _____ earn money from illegal drugs.

5. Two illegal drugs are cocaine and _____.

6. Subsistence farmers in Peru grow a plant called _____.

Group B

A. empire

B. marijuana

C. Quechua

D. drug traffickers

E. terrorists

F. coca

◆ Read and Remember

Write the Answer   Write one or more sentences to answer each question.

1. What mountain chain is found along the west coast of South America?

2. What are the four climate regions of the Andes Mountains?

3. Look at the map on page 82. What are the five Andean countries?

4. What have terrorists done in Peru?

5. What does the United States buy from Chile in the winter?

6. What is the most important cash crop in Colombia?

Find the Answer   Find the sentences that tell about a problem in the Andean countries. On your paper write the sentences you find. You should find three sentences.

1. For thousands of years, Indians have lived in Peru.

2. Drug traffickers sell illegal drugs from Peru and Colombia.

3. It is very hard to build roads through the Andes Mountains.

4. Fruits and vegetables grown in Chile are sold in the United States.

5. Terrorists have tried to win control of Peru's government.

6. Colombia earns money from exporting coffee and oil.

◆ Think and Apply

Categories   Read the words in each group. Decide how they are alike. Find the best title for each group from the words in dark print. Write the title on your paper.

Peru   Climates in the Andes   Problems in the Andean Countries
Chile   Colombia

1. has borders on the Pacific Ocean
        and the Caribbean Sea
    coffee is the main cash crop
    has many drug traffickers

2. Inca empire
    largest Indian population
    capital is on the Pacific coast

3. hot land at sea level
    cold land at 10,000 feet
    frozen land above 15,000 feet

4. about 2,000 miles long
    cold climate in the south
    about 100 miles wide

5. poverty
    terrorism
    drug traffickers

◆ Journal Writing

Write a paragraph that tells five facts about the Andean countries.

Using Latitude and Longitude

You learned that lines of **latitude** are east and west around Earth. Lines of **longitude** are north and south lines. Lines of latitude and longitude form a grid on maps and globes. These grids make it easy to find places.

The latitude of a place is written first, and the longitude is written next to it. Lima, Peru, is close to 12°S latitude. It is near 77°W longitude. We say that the latitude and longitude of Lima is 12°S, 77°W.

Look at the map below of the Andean countries. On your paper write the answer that completes each sentence.

1. The latitude and longitude of Bogota, Colombia, is _____.

   60°N, 12°E   5°N, 74°W   80°S, 30°E

2. The city with the latitude and longitude of 0°, 78°W is _____.

   Quito   Bogota   Santiago

3. The city with the latitude and longitude of 16°S, 68°W is _____.

   Punta Arenas   La Paz   Quito

4. The latitude and longitude of Santiago, Chile, is _____.

   15°N, 105°E   75°N, 123°E

   33°S, 71°W

5. The city with the latitude and longitude of 53°S, 71°W is _____.

   Lima   Cali   Punta Arenas

CHAPTER 11

Understanding Latin America

Think About As You Read

1. Why are poverty and rapid population growth problems in Latin America?
2. How does the lack of transportation hurt Latin America?
3. What problems can a nation have with a one-crop economy?

New Words

◆ land reform
◆ rapid population growth
◆ illiteracy
◆ one-crop economy
◆ communism

People and Places

◆ Pan American Highway

Mr. Sanchez grows coffee on a large plantation in Colombia. There are hundreds of workers on his plantation. Mr. Sanchez is rich because he earns a lot of money by exporting his coffee. But his plantation workers are very poor. They earn low salaries. They struggle to have enough food for their families. The difference between the rich and the poor is one of many problems in Latin America. In this chapter you will learn about six problems in this region. You will find out how Latin Americans are working to solve these problems.

Poverty and Rapid Population Growth

The first problem is poverty. Millions of people in Latin America are very poor. Many farmers are poor because they do not own enough land to grow cash crops. Subsistence farmers struggle to grow enough to feed their families. Millions of people are poor because they do not have jobs. Other people have jobs that pay

About one third of the people in Latin America are under the age of 15 because of the region's rapid population growth.

very low salaries. Many people have moved to the United States to find jobs and earn more money.

Every day, poor people move to the large cities of Latin America. There they hope to find better jobs. They live in slums that surround the cities. The cities are growing fast. They are becoming crowded. The cities do not have enough schools, hospitals, and services for all the people in the slums.

Latin American governments are working to end poverty. Some governments are building places in the cities where poor families can live. Some governments are using **land reform**. Land reform happens when the government buys large plantations. Then they divide the land into smaller farms. These farms are given to poor farmers. Land reform helps farmers have their own land. However, farmers also need better tools and seeds in order to grow larger crops.

A second problem in Latin America is **rapid population growth**. Rapid population growth means the population grows a lot each year. For example, in 1985 Mexico had 80 million people. In 1996 there were 96 million people. Many people believe the population of Latin America will double in 20 years. Rapid population growth is a problem because there are not enough jobs, food, and homes for all the people.

Millions of people live in poverty in Latin America.

Many farmers in Latin America are poor. Children often have to work on their family's farm instead of going to school.

There are two reasons why the population is growing fast. One reason is that Latin Americans like big families. Farmers want to have children who will help with the farm work. A second reason is that there is better health care today. Fewer young babies die. People are healthier and live longer. In some places people are trying to have smaller families.

Illiteracy and Problems Using Resources

The third problem is **illiteracy**. Illiteracy means people do not know how to read and write. For example, in Guatemala almost half of the people cannot read. In El Salvador about one fourth of the people cannot read. People who cannot read cannot get good jobs.

Schoolchildren in Mexico

In many Latin American countries, young children go to school. But they are too poor to go to high school. Instead they work to help their families. In mountain villages and rain forest towns there are fewer schools and teachers. Children are not able to learn science, math, and computer skills. Without these skills, they will always work at low-paying jobs. To solve the education problem, many countries are building more new schools.

The fourth problem is the lack of good roads and transportation. Latin America is rich in resources. But many resources are in deserts and rain forests. Other resources are in mountains. There are not enough roads through deserts, rain forests, and mountains. So the people of the region cannot use many of their resources. To solve this problem, governments must build more roads and railroads.

To help trade, the countries of Latin America have built the Pan American Highway. The Highway includes about 30,000 miles of roads. These roads connect 17 countries in Latin America. But these roads are still not enough.

The Pan American Highway

One Cash Crop Is Not Enough

The fifth problem is that too many countries depend on one cash crop to earn money. Colombia earns most of its money from exporting coffee. Most of the money Cuba earns is from sugarcane. Countries in Latin America sell some other cash crops, too. But most of their money comes from one main crop.

Countries that earn most of their money from one crop have a **one-crop economy**. A one-crop economy can be a problem. What happens when there is little rain? What happens when the sugarcane crop or the coffee crop is not very large? What happens when the prices for these crops go down? Then these nations earn very little money.

To solve this problem, nations are planting different kinds of crops. Nations also need more factories and industries. They need to make many kinds of factory products that can be exported.

Many Latin American countries earn money from tourism. This helps the country change from a one-crop economy.

These elections in Nicaragua show that democracy is spreading in Latin America.

Changes in Latin America are making life better for the people living there.

The Need for Peace and Democracy

The sixth problem is civil wars and terrorism. You have learned how civil wars hurt Central America. In Chapter 10 you learned how terrorism has hurt Peru. In Mexico, Colombia, Peru, and some other countries, drug traffickers have their own soldiers. They attack and kill people who try to stop the illegal drug trade. In order for Latin America to become more developed, the region needs peace. All nations must work to stop terrorism. They need to end civil wars. The illegal drug trade must stop.

For many years people feared that Communists would win control of governments everywhere. Today **communism** has lost most of its power. Under communism the government controls a country's businesses. Today in Latin America only Cuba has a Communist government.

Democracy is spreading in Latin America. Countries such as Brazil, Peru, Chile, and Colombia were once ruled by dictators. They are now working to have democratic governments. People vote for their leaders in free elections. Most people are enjoying more freedom today.

Latin America has developed in many ways. Its people have better schools, better health care, and more jobs. With peace and better use of resources, Latin America will have more developed nations.

Chapter Main Ideas

1. Land reform is helping to end poverty in Latin America. Cities need better places for poor families to live.
2. The people of Latin America need to build better roads and transportation. Then they will be able to use the region's natural resources.
3. There are fewer dictators in Latin America today. Many nations are trying to be democracies.

◆ Vocabulary

Find the Meaning Choose the word or words that best complete each sentence. Write your answers on your paper.

1. **Land reform** means breaking up _____ and giving land to poor farmers.

   ball fields   parks   plantations

2. Under **communism**, businesses are owned by the _____.

   rich people   government   students

3. **Illiteracy** means people do not know how to _____.

   paint   farm   read

4. If a country has **rapid population growth**, then it has more and more _____ each year.

   farms   people   cash crops

5. A country with a **one-crop economy** has one main _____.

   cash crop   subsistence crop   factory product

◆ Read and Remember

Finish the Paragraph Number your paper from 1 to 5. Use the words in dark print to finish the paragraph below. Write the words you choose on your paper.

high school   lower class   democratic   terrorists   peace

Latin America has a very small upper class and a very large ___1___. So there are millions of poor people. Many poor children go to work instead of to ___2___. Many people have been killed by ___3___, civil wars, and drug traffickers. Many countries want to have ___4___ governments. The countries of Latin America want to become more developed. They need more schools, better health care, and more jobs. Most of all, Latin America needs ___5___.

Find the Main Idea   On your paper copy the boxes shown below. Then read the five sentences. Choose the main idea and write it on your paper in the main idea box. Then find three sentences that support the main idea. Write them on your paper in the boxes of the main idea chart. There will be one sentence in the group that you will not use.

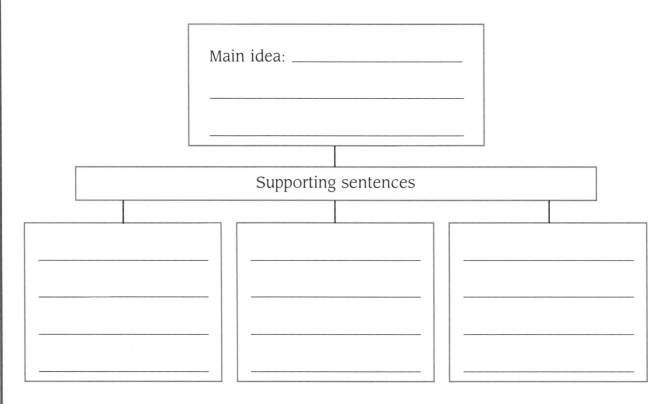

a. The population is growing too fast.

b. Latin America must solve many problems for its nations to become developed.

c. Better roads and railroads are needed.

d. Chile exports fruit to the United States.

e. Illiteracy is a problem in some nations.

◆ **Journal Writing**

You have read about the nations of Latin America. These nations are different. They have different landforms, climates, and governments. But they all have problems. Choose two problems that you think Latin America must solve. Write a paragraph that tells why nations must solve these problems.

Western Europe

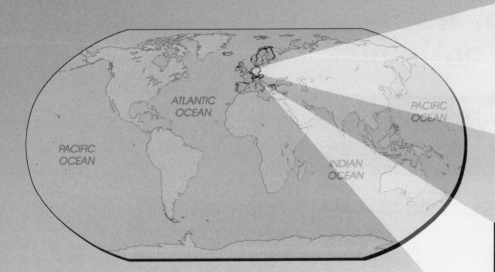

London, England

Venice, Italy

DID YOU KNOW?

▲ The first French fries were made in Belgium, not in France.

▲ Amsterdam, the capital of the Netherlands, was built on land that was once below the sea.

▲ There is no speed limit on some highways in Germany.

▲ The city of Venice, Italy, has about 400 bridges.

▲ Inside Italy is the world's smallest country, Vatican City. Inside this tiny country is the world's largest palace, the Vatican Palace. It has more than 1,000 rooms.

WRITE A TRAVELOGUE

For hundreds of years, the people of Western Europe spread their ideas and way of life to other parts of the world. Look through the photographs in this unit. What do you see in Western Europe that looks like where you live? Write a paragraph that describes what you see. Your travelogue should explain how the place is similar to where you live. After you read the unit, write two paragraphs that explain how two ideas have moved from Western Europe to other parts of the world.

THEME: MOVEMENT

CHAPTER 12

Western Europe: The Western Nations of a Small Continent

Where Can You Find?

Where can you find a small country with three official languages?

Think About As You Read

1. How have the seas helped Western Europe?
2. What types of government are found in Western Europe?
3. How are the countries in Western Europe different from each other? How are they alike?

New Words

- peninsula
- raw materials
- ocean currents
- fertile
- constitutional monarchy
- steel
- agriculture
- fertilizer

People and Places

- Western Europe
- Italy
- Germany
- North Sea
- Baltic Sea
- Mediterranean Sea
- North European Plain
- Alps
- Rhine River
- London
- Paris
- Christians
- Jews
- Muslims
- Switzerland

An American decided to buy a car made in Europe. Several countries in Western Europe make cars. Which country did the new car come from? It came from Italy. Cars are also made in Great Britain, Germany, France, and other countries in Western Europe. Cars are one of many products that are made in this region.

Geography and Climate of Western Europe

Europe is the second smallest continent. The region called Western Europe includes the countries shown on the map on page 97. Western Europe is only one third the size of the United States. But this region has many countries and more than 380 million people. It has far more people than the United States. This region has a high population density.

Western Europe is a large **peninsula**. A peninsula is land that has water on three sides. Many countries in

WESTERN EUROPE

MAP KEY
- ⊛ Capital city
- —·— Border between countries
- ⌇ River

ICELAND

ATLANTIC OCEAN

SWEDEN
FINLAND
NORWAY

GREAT BRITAIN
IRELAND
North Sea
DENMARK
Baltic Sea

London ⊛
NORTH EUROPEAN PLAIN
GERMANY

Paris ⊛
Rhine R.
FRANCE
SWITZERLAND
ITALY

PORTUGAL
SPAIN
GREECE
CYPRUS

Mediterranean Sea

There are many countries in Western Europe. Which small country is located between France and Italy?

Western Europe are smaller peninsulas. Look at the map. Which countries in this region are peninsulas?

The seas are very important in Western Europe. No place in this region is more than 300 miles from a sea. The Atlantic Ocean is to the west. The Arctic Ocean, the North Sea, and the Baltic Sea are in the north. The Mediterranean Sea is to the south.

Europe has a long coast. It has many good ports. For hundreds of years, the seas have helped Europe. People from many countries in Europe have used the seas to sail to far-off lands. For hundreds of years, European countries ruled many colonies in Asia, Africa, and the Americas.

The seas have helped Europe in three other ways. First, people use the seas for fishing. Second, Europeans use the seas for shipping and trading. Trading ships carry goods and **raw materials** to Europe from many lands. Raw materials are natural resources that are used to make products. Countries in Western Europe make products from the raw materials. Then they sell the products to many nations.

A car factory in Italy

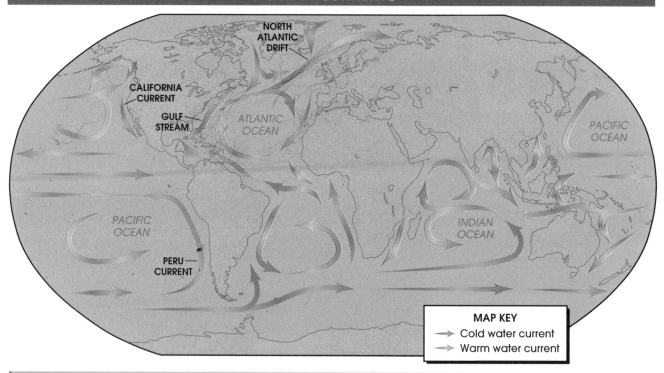

CALIFORNIA CURRENT
GULF STREAM
ATLANTIC OCEAN
PACIFIC OCEAN
NORTH ATLANTIC DRIFT
PACIFIC OCEAN
PERU CURRENT
INDIAN OCEAN

MAP KEY
→ Cold water current
→ Warm water current

Warm water and cold water currents affect the weather around the world. Which two currents bring warm water in the direction of Western Europe?

Skiers in the Alps

Third, the seas help Western Europe by giving the region a mild climate. Did you know that Western Europe is as far north as Canada? But Western Europe has a much warmer climate. This happens because the warm **ocean currents** bring warmer air to most of Western Europe. Ocean currents are streams of water in the ocean. Ocean currents can be warm water or cold water. Warm ocean currents cross the Atlantic Ocean from the Americas to Western Europe. These warm ocean currents keep the region's climate mild.

Plains and mountains are the two main landforms in Western Europe. The North European Plain covers a large part of Western Europe. This large flat area stretches across the northern part of Europe. The soil on the plain is **fertile**. This land is very good for farming. Mountains cover the most northern parts of Europe. There are also mountain chains in parts of southern and central Europe. The Alps are Europe's most famous mountain chain. Many mountains in the Alps are covered with snow all year.

There are many different cultures in Western Europe. These dancers are from Italy.

Western Europe has many important rivers. This region also has many canals. The longest river in Western Europe is the Rhine River. The Rhine River is 820 miles long. It is much shorter than the Mississippi and Amazon rivers. The rivers and canals of Western Europe form important waterways. There are many port cities along these waterways. Ships carry goods from these ports to the seas. Then the goods are shipped to other countries.

Cities, People, Cultures, and Government

Most people in Western Europe live in cities. Most large cities in this region are located on rivers. Cities like London, England, and Paris, France, are famous.

Who are the people of Western Europe? Most Europeans are white. But many people from Africa and Asia have moved to Europe in order to find jobs.

Most people in Western Europe are Christians. They belong to Roman Catholic or Protestant churches. Jews and Muslims are smaller religious groups found in Western Europe.

Each country in Western Europe has its own money.

There are many different countries in Western Europe. Each one has its own laws, stamps, and flags. Each country has its own culture with special foods, customs, and holidays. Almost every country has its own language. In some countries a few different languages are spoken. Switzerland is a small country

The Rhine River

with three official languages. Most Europeans can speak two or more languages. Most children in Western Europe spend many years in school. Almost everyone knows how to read and write.

Most countries in this region are democracies. Some countries also have a king or a queen. The king or queen has little power. Instead a president or a prime minister leads the country. A democracy that has a king or a queen is a **constitutional monarchy**.

Natural Resources and Earning a Living

Western Europe is a region with developed nations. Most people have a high standard of living. The region has many factories and many industries.

Western Europe is not rich in natural resources. But it has some resources that are used in factories. Waterpower from rivers is used to make electricity. Many countries have coal and iron. Coal and iron are used to make **steel**. Some countries, such as Great Britain, have oil.

Many people in Europe work at service jobs. Others work at factory jobs. A small part of the population works at **agriculture**, or farming. Europeans use modern farm machines and lots of **fertilizer**. Fertilizers make soil more fertile to help grow more crops. Most countries grow enough food for their people. Some countries export food.

Train station in Western Europe

There is excellent public transportation throughout Western Europe. People can easily travel within their own country or from one country to another by train.

Four large countries in Western Europe are often in the news. These leading countries are Great Britain, France, Germany, and Italy. In the next chapters, you will learn why these countries are important.

Chapter Main Ideas

1. Western Europe is a large peninsula. Many countries in the region are smaller peninsulas.
2. Most of the countries of Western Europe are democracies. Some are constitutional monarchies.
3. There are many countries, cultures, and languages in Western Europe.

◆ Vocabulary

Finish Up   Choose the word or words in dark print that best complete each sentence. Write the word or words on your paper.

steel   fertilizer   ocean currents   peninsula   agriculture

1. A _____ is land that has water on three sides.

2. Streams of water in the sea are called _____.

3. Coal and iron are used to make _____.

4. Another word for farming is _____.

5. To grow more crops, farmers improve the soil by adding _____.

◆ Read and Remember

Find the Answer   Find the sentences that tell why Western Europe is a region with developed nations. Write the sentences you find on your paper. You should find five sentences.

1. Farmers grow enough food for the region.

2. Western Europe has many factories.

3. Each country has its own laws, stamps, and flag.

4. Almost everyone can read and write.

5. People speak many different languages in Western Europe.

6. Ships bring to Europe raw materials that people use to make factory products.

7. European ships carry factory products to many countries of the world.

8. Europe is the second smallest continent.

9. Western Europe has a long coast.

10. Ocean currents bring warm air to most of Western Europe.

Drawing Conclusions   Read each pair of sentences. Then look in the box for the conclusion you might make. Write the letter of the conclusion on your paper.

1. People use the seas for fishing.
   People use the seas for trading and shipping.

   Conclusion: _____

2. Western Europe is warmer than Canada.
   Western Europe's winters are not too cold, and summers are not too hot.

   Conclusion: _____

3. There are port cities on many rivers.
   People travel on rivers through many parts of Western Europe.

   Conclusion: _____

4. Western Europe has coal and iron.
   Western Europe has waterpower and some oil.

   Conclusion: _____

5. Each country in Western Europe has its own leaders, government, and money.
   Each country has its own culture and flag.

   Conclusion: _____

Conclusions
   A. Rivers are important in Western Europe.
   B. Western Europe has some of the resources it needs for its factories.
   C. Warm ocean currents give Western Europe a mild climate.
   D. The countries of Western Europe have different cultures.
   E. The seas help Western Europe.

◆ **Journal Writing**

Write a paragraph that tells why many people in Western Europe have a high standard of living.

CHAPTER 13

The United Kingdom: An Island Country

Where Can You Find?
Where can you find Great Britain's oil?

Think About As You Read

1. What is the United Kingdom?
2. How have the seas helped the United Kingdom?
3. How did the Industrial Revolution change the United Kingdom?

New Words

◆ petroleum
◆ natural gas
◆ Act of Union
◆ Industrial Revolution
◆ revolution
◆ ethnic groups
◆ monarchs
◆ world power

People and Places

◆ United Kingdom
◆ Queen Elizabeth
◆ English Channel
◆ England
◆ Scotland
◆ Wales
◆ Northern Ireland
◆ Ireland
◆ Thames River
◆ Buckingham Palace
◆ British Empire
◆ Channel Tunnel
◆ Folkestone
◆ Calais

Canada and Jamaica have a queen, but she does not live in those countries. Her home is in the United Kingdom. Queen Elizabeth is the queen of Canada, Jamaica, the United Kingdom, and many other countries. In this chapter you will read about the United Kingdom, the home of Queen Elizabeth.

What Is the United Kingdom?

The United Kingdom is a country that is separated from the rest of Europe by the English Channel. The country has several islands. Great Britain is the largest island of the United Kingdom. Sometimes the entire country is called Great Britain. England, Scotland, and Wales are on the island of Great Britain. Northern Ireland is the fourth part of the United Kingdom. It is on the northern part of the island of Ireland. The United Kingdom's full name is the United Kingdom of Great Britain and Northern Ireland.

The people of the United Kingdom are often called the British. English is their official language. More than

The United Kingdom is an island country. What separates this country from the rest of Europe?

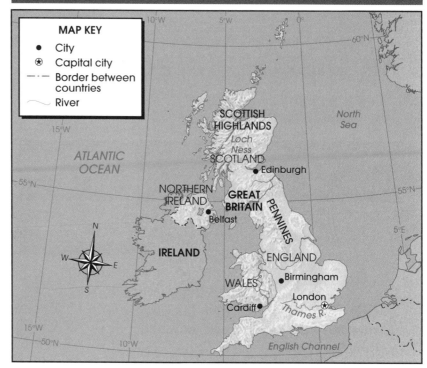

Flag of the United Kingdom

A lake in the mountains of Scotland

58 million people live in the United Kingdom. Most of these people live in England. England is the largest part of the United Kingdom.

Landforms, Seas, and Cities

The United Kingdom has a variety of landforms. Mountains cover large parts of Scotland and Wales. Hills cover much of Northern Ireland and northern England. Plains cover southern England. Southern England has the best farmland in the United Kingdom.

Seas surround the United Kingdom. The Atlantic Ocean is to the west. The North Sea is to the east. The English Channel is to the south. Seas are important to the United Kingdom. No place in this country is more than 75 miles from a sea. Warm ocean currents give the country a mild climate. The currents also bring much rain. The country has a long coast with many good harbors. So the British do a lot of shipping and trading. The British also catch and eat plenty of fish.

The Thames River is the country's busiest waterway. It is in southern England. Ships on the Thames River carry British goods to the English Channel. From there British ships sail to many countries.

Tourists in London enjoy a visit to Buckingham Palace.

Most British people live in cities. London, the country's capital, is the largest city. It is a busy port on the Thames River. Tourists enjoy visiting London's museums, stores, and theaters. In the summer they can visit the queen's home at Buckingham Palace.

Natural Resources and Earning a Living

The United Kingdom is not rich in natural resources. Its most important resource is oil. Oil is also called **petroleum**. Britain's oil is in the ground under the North Sea. Britain also has **natural gas** and coal. Natural gas is a type of fuel. Coal is mined in England and Wales. Great Britain has enough oil, natural gas, and coal for its own needs. It also exports some oil.

Most British people work at service jobs. Many work in factories. Still others work at shipping and trading. Most people enjoy a high standard of living.

A small part of the population works at farming. Britain's main crops are wheat and other grains. Most British farmers also raise sheep. The British are good farmers. But their climate is not hot enough to grow such crops as bananas, oranges, and pineapples. So the British import about one third of their food.

The British take oil from the ground under the North Sea.

The United Kingdom's History

Long ago, England, Wales, Scotland, and Ireland were independent countries. In 1707 a law was passed

Queen Elizabeth

called the **Act of Union**. With that law England, Wales, and Scotland became one nation.

In 1801 another Act of Union added Ireland to Great Britain. But many Catholics in Ireland did not want to be part of Great Britain. In 1921 the southern part of Ireland became a separate country. Northern Ireland remained part of the United Kingdom.

During the 1700s the **Industrial Revolution** began in England. This **revolution** was a change from making products by hand in homes to making products by using machines in factories. The United Kingdom became the world's first industrial nation. It used its coal and iron in its new factories. It used the seas to ship factory products to many countries.

During the 1700s the British began building a huge empire. By 1900 the British Empire had colonies in every part of the world. Great Britain got raw materials for its factories from these colonies. After 1945 most British colonies became free.

Today 50 countries that had been part of the British Empire are members of the British Commonwealth. Queen Elizabeth is the queen of the United Kingdom and 15 other Commonwealth countries.

Immigrants have moved to the United Kingdom from many Commonwealth countries. Today there are many **ethnic groups** in the United Kingdom. Many Africans and Asians in Great Britain do not live as well as other British people. The government is trying to find ways to help all people.

The Government of the United Kingdom

The United Kingdom is a constitutional monarchy. In a constitutional monarchy, the king or queen has very little power. Long ago, England's kings and queens, or **monarchs**, made all laws. Slowly, over hundreds of years, these monarchs lost their power. Parliament, the country's lawmakers, slowly grew more powerful. Today Parliament makes all laws. Queen Elizabeth does not make laws. She is a symbol of the nation. Most British people love their queen.

People of different ethnic groups from around the world live in the United Kingdom.

British soldiers have been in Northern Ireland for many years.

The British are proud of their government. It was the first democracy in Western Europe. The British vote for people to make laws for them in Parliament. The prime minister is the leader of Parliament and of the entire country.

Looking at the Future

In the late 1960s, fighting began between Catholics and Protestants in Northern Ireland. Many British soldiers were sent to Northern Ireland. The fighting in Northern Ireland has lasted more than thirty years. In 1994 the fighting stopped for a while. Will peace come to Northern Ireland? Will Northern Ireland remain part of the United Kingdom? No one knows these answers.

The United Kingdom no longer rules a huge empire. But the United Kingdom continues to be a **world power**. The British are proud that their small nation is an important leader among the countries of the world.

Parliament building in London

Chapter Main Ideas

1. England, Scotland, Wales, and Northern Ireland form the United Kingdom.
2. The United Kingdom used the seas to build a huge empire. Today it uses the seas for trade.
3. Great Britain became the world's first industrial nation. It got raw materials for its factories from colonies in the empire.

The Channel Tunnel

For hundreds of years, people dreamed of building a tunnel under the English Channel. That dream came true in 1994. After six years of work, the Channel Tunnel was finished! Trains began traveling through the tunnel between England and France. The Channel Tunnel, also called the Chunnel, was built at the most narrow part of the English Channel. The tunnel joins the city of Folkestone, England, with the city of Calais, France.

The Channel Tunnel is in the region of Western Europe. The region's technology and industry made it possible to build the tunnel. It cost $15 billion to build.

Today the Channel Tunnel makes it possible to travel between London and Paris in about three hours. The train ride through the Channel Tunnel takes only 20 minutes. Only trains can travel through the tunnel. But special trains can carry cars and trucks through the Channel Tunnel.

For hundreds of years, people needed ships to cross the English Channel. Today people can cross the English Channel by train through the Channel Tunnel.

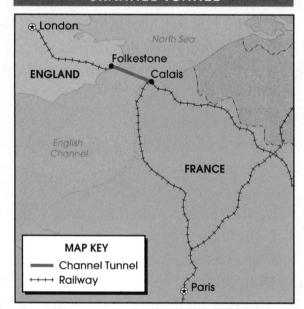

CHANNEL TUNNEL

MAP KEY
— Channel Tunnel
+++++ Railway

Write a sentence to answer each question.

1. **Place**   How is the Channel Tunnel different from every other place?

2. **Human/Environment Interaction**   How has the Channel Tunnel changed the United Kingdom and France?

3. **Region**   In what kind of region is the Channel Tunnel?

4. **Location**   Look at the map. Where was the Channel Tunnel built?

5. **Movement**   How do cars and trucks travel through the Channel Tunnel?

◆ Vocabulary

Finish the Paragraph   On your paper write the words in dark print that finish the paragraph below.

monarch   **Act of Union**   **Industrial Revolution**   **petroleum**

England, Scotland, and Wales became one country in 1707 when a law called the __1__ was passed. During the 1700s England was the first country to become a nation where products were made by machines in factories. This change was called the __2__. In England's democracy the queen, or __3__, has little power. The United Kingdom now gets oil, or __4__, from the North Sea.

◆ Read and Remember

Matching   Each item in Group B tells about an item in Group A. Write the letter of each item in Group B next to the correct number on your paper.

Group A

1. Northern Ireland

2. Parliament

3. English Channel

4. Commonwealth

5. Great Britain

6. London

Group B

A. This is the group that makes laws in the United Kingdom.

B. This is another name for the United Kingdom.

C. In this part of the United Kingdom, Catholics and Protestants have fought for many years.

D. This capital city is on the Thames River.

E. This is a group of nations that were part of the British Empire.

F. This body of water separates the United Kingdom from the continent of Europe.

◆ Think and Apply

Find the Main Idea   On your paper copy the boxes shown below. Read the five sentences. Choose the main idea and write it on your paper in the main idea box. Then find three sentences that support the main idea. Write them on your paper in the boxes of the main idea chart. There will be one sentence in the group that you will not use.

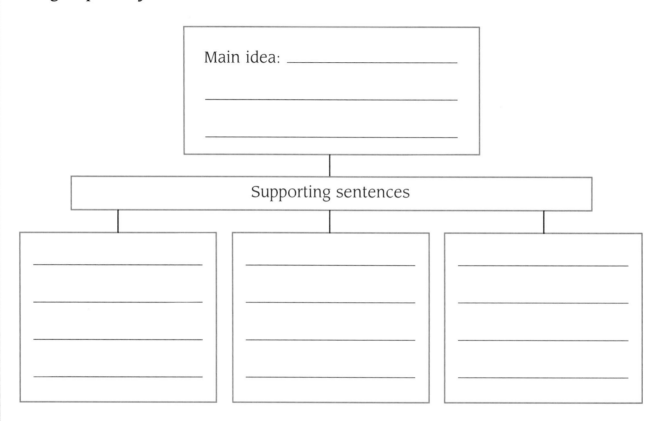

a. The British used the seas for trading.

b. Mountains cover large parts of Scotland and Wales.

c. The seas have helped Britain in many ways.

d. The British catch and eat a lot of fish.

e. The British used the seas to build an empire.

◆ Journal Writing

Journal Writing   Imagine you are telling a friend to visit the United Kingdom. Write a paragraph with three or more reasons to visit the United Kingdom.

Reading a Bar Graph

Special drawings called **graphs** can help us compare facts. The graph below is called a **bar graph**. It shows facts by using bars of different lengths. This bar graph shows populations. By looking at the length of the bars, you can tell the population of each part of the United Kingdom. You can also compare the populations of the four different parts of the country.

Look at the bar graph below. Write on your paper the answer to each question.

1. What is the population of Northern Ireland?

15,000,000   1,640,000   7,500,000

2. What is the population of Wales?

2,900,000   12,300,000   35,200,000

3. What is the population of England?

48,700,000   30,420,000   6,400,000

4. What is the population of Scotland?

42,900,000   26,300,000   5,130,000

Write on your paper the answer to each question below.

1. Which country has the largest population?

2. Which country has the smallest population?

3. What conclusion can you draw about the population of the four parts of the United Kingdom?

4. What two parts of the United Kingdom have a larger population than Wales?

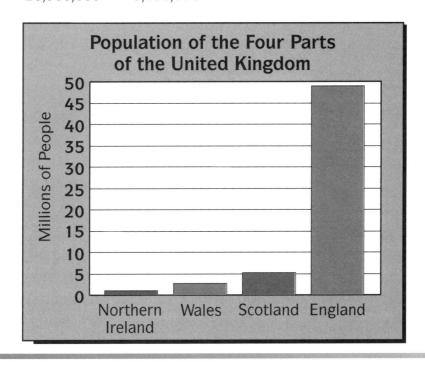

Population of the Four Parts of the United Kingdom

France: A Leader of Farming and Industry

Where Can You Find?
Where can you find the tallest mountain in Western Europe?

Think About As You Read

1. What kinds of landforms and climates are in France?
2. How do France's resources help its farms and factories?
3. What is special about France's culture?

New Words

- Mediterranean climate
- bullet trains
- uranium
- nuclear energy
- nuclear power plants
- manufacture
- cafés
- universities
- unemployment

People and Places

- Corsica
- Pyrenees Mountains
- Mont Blanc
- French Riviera
- Seine River
- Marseilles
- Eiffel Tower
- Algeria

Two American friends want to visit a country in Western Europe. One friend wants to be in northern Europe. She wants to see a big city with museums. The other friend wants to visit southern Europe. She wants to swim at a beach on the Mediterranean Sea. They can visit the same country if they visit France.

Landforms and Climate

France has the largest land area of any country in Western Europe. Find France on the map on page 113. The Atlantic Ocean is to the west. The English Channel is to the north. The Mediterranean Sea is to the south. The island of Corsica belongs to France. It is in the Mediterranean Sea.

Like the United Kingdom, the northern part of France has a mild climate. Southern France has a **Mediterranean climate**. This climate has long, hot, dry summers and short, rainy winters. Many nations on the Mediterranean Sea have this type of climate.

FRANCE

France has several different landforms, including plains and mountains. Which mountains are located between France and Spain?

France's flag

Plains cover more than half of France. The North European Plain covers northern France. Paris, the capital of France, is on this plain. This region has many farms and industries. Plateaus cover the northeastern part of the country.

In the southwest of France, the Pyrenees Mountains separate France from Spain. In the east the Alps Mountains separate France from Italy and Switzerland. One mountain in the Alps, Mont Blanc, is the tallest mountain in Western Europe.

The famous French Riviera is in the southeast. It has beautiful beaches on the Mediterranean Sea. Many tourists from around the world visit the Riviera.

Movement: Transportation in France

There are five important rivers in France. Paris was built on the Seine River. The Rhine River forms part of France's border with Germany. Canals join the rivers together. Ships can travel from the North Sea down the rivers and canals of France to the city of Marseilles. Marseilles is a city on the Mediterranean Sea. It is France's busiest port.

Mont Blanc

The waterways are just one type of transportation in France. Most families have their own cars. The country has excellent highways. Paris has subway trains. Railroad trains run throughout the country. France has some of the world's fastest trains. These **bullet trains** travel faster than 180 miles an hour. They go from Paris to other French cities. Some of these bullet trains go to other countries in Europe.

Natural Resources, Industries, Farms, and Cities

France is a rich country with many farms and many industries. Most people enjoy a high standard of living. But France was not always a rich country. When World War II ended in 1945, much of the country had been destroyed. The United States helped France rebuild its cities and factories.

France has some of the natural resources it needs to be an industrial country. It has coal and iron. It has bauxite for making aluminum. France also has a mineral called **uranium**. Uranium is used to make **nuclear energy**. France has little oil and natural gas. So the French make three fourths of their electricity in **nuclear power plants**.

France is an industrial leader. The French **manufacture**, or make, steel, planes, and weapons. They make fine perfumes and clothing. Computers,

Bullet train

France gets most of its electricity from nuclear power plants like this one in the Rhone River Valley.

Both tourists and people who live in Paris enjoy meals in the city's many sidewalk cafés.

chemicals, trains, and cars are some other French products. Only the United States, Japan, and Germany make more cars than France. French factory products are sold to many countries.

France also exports many foods. France has fertile soil for farming. Only a small part of the population works in agriculture. But the French grow more food than any other country in Western Europe. The French are famous for their cheese and wine. They make more than 300 different kinds of cheese. They grow many kinds of grapes. They use these grapes to make different types of wine.

The French and Their Culture

France has 58 million people. Three fourths of the French people live in cities. Paris is the largest city. About 9 million people live in and around the capital. Paris has many factories. Millions of tourists visit Paris. They eat in its outdoor **cafés**, or restaurants. They enjoy art museums and boat rides on the Seine River. One of the most famous places to see is the Eiffel Tower. From the top of the Eiffel Tower, tourists can view Paris and the Seine River.

France is a democracy. The people of France vote for leaders to make their laws in France's Parliament. Every seven years the people of France vote for their president.

The Eiffel Tower near the Seine River in Paris

The French Riviera

The French have a special culture. Many of the world's greatest artists and writers have been French. The French people love their French language. They try not to speak other languages.

Cooking is a type of art for the French. Meals have to taste good and look good. The French make special meals using different types of sauces. People enjoy eating crisp French bread and hot onion soup with their meals. Desserts are also special.

Enjoying life is another part of French culture. All workers get five weeks of vacation time during the summer. They use vacation time to go to the beach, to travel, and to go camping. They enjoy sitting in a café while watching people walk by on the sidewalk.

Education is important to the French. All children must go to school for ten years. Many students study in **universities** after they finish high school.

French desserts

Who are the people of France? Most French are white people. They come from families that have lived in Europe for a long time. Most people are Catholics. But there are also Muslims and Jews. About 4 million people are immigrants from southern Europe, northern Africa, and Asia. Many immigrants come from countries like Algeria that were once ruled by France. Most immigrants do not live as well as other French people. Many immigrants do not have jobs. **Unemployment** is a problem in France. More than ten percent of France's people do not have jobs.

The French are proud people. They love their language and their culture. They are proud of their many farms and factories. They are proud that France is a world power. France is a leader in Europe and in the world today.

Chapter Main Ideas
1. France is a leader of industry and farming in Western Europe. It has a high standard of living.
2. France has excellent transportation because of its highways, trains, and waterways.
3. The French are proud of their language and their special culture.

◆ Vocabulary

Analogies   Use the words in dark print that best complete the sentences. Write the correct answers on your paper.

Mediterranean   **manufacturing**   **unemployment**
uranium   **university**

1. Restaurant is to café as school is to _____.

2. Farming is to crops as _____ is to factory products.

3. Tropical climate is to Brazil as _____ climate is to Riviera.

4. Bauxite is to aluminum as _____ is to nuclear energy.

5. Not being able to read is to illiteracy as not having a job is to _____.

◆ Read and Remember

Complete the Geography Organizer   On your paper copy the geography organizer and complete it with information about France.

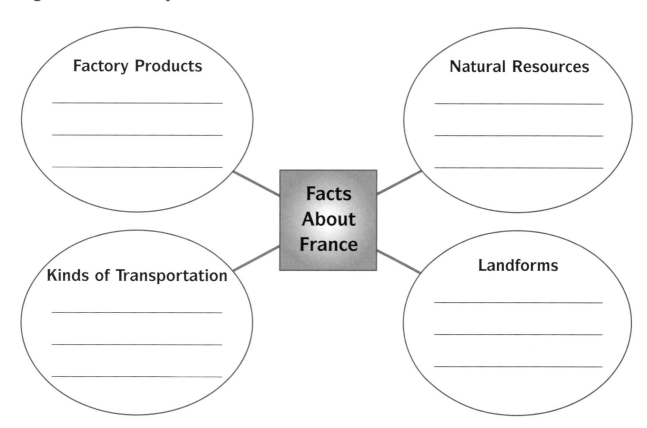

Write the Answer   Write one or more sentences to answer each question.

1. What type of weather is found in a Mediterranean climate?

2. What is the busiest port in France?

3. What types of trains are used in France?

4. Where is nuclear energy made?

5. What landform covers more than half of France?

6. What island in the Mediterranean Sea belongs to France?

7. The Pyrennees Mountains separate France from what other country?

8. From what source of energy does France get most of its electricity?

◆ Think and Apply

Fact or Opinion   Number your paper from 1 to 8. Write **F** on your paper for each fact. Write **O** for each opinion. You should find three sentences that are opinions.

1. There are mountains in the eastern part of France.

2. France grows more food than any other nation in Western Europe.

3. The French make many kinds of wine and cheese.

4. The French Riviera has the best beaches in the world.

5. The French manufacture cars, trains, planes, and clothing.

6. Paris is the most beautiful city in the world.

7. Many tourists visit the Eiffel Tower.

8. France has too many immigrants.

◆ Journal Writing

Imagine you are visiting France. Write a paragraph that tells what you would do in France. Tell about four or more places you would visit.

Understanding Circle Graphs

A **circle graph** is a circle that has been divided into parts. Each part looks like a piece of pie. All the parts make up the whole circle. All the parts equal 100 **percent**.

The circle graph on this page shows how people earn a living in France. The three groups of workers make up the whole working population, or work force, in France.

Look at the circle graph. Choose the words in dark print to complete the sentences below. Write the words you choose on your paper.

7   31   62   service jobs   farming   developed

1. People who work in industry are _____ percent of the work force.

2. People who work in agriculture are _____ percent of the work force.

3. People who work at service jobs are _____ percent of the work force.

4. The largest group in the French work force does _____.

5. The smallest group in the work force does _____.

6. The ways people in France earn a living tell us that France is a _____ nation.

The Work Force in France

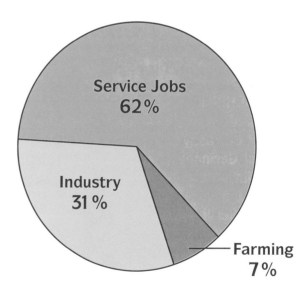

Service Jobs
62%

Industry
31%

Farming
7%

CHAPTER 15

Germany: One United Country

Think About As You Read

1. Why was Germany divided into two countries after World War II?
2. How did Germany become a united country again?
3. What problems does Germany have today?

New Words

◆ defeated
◆ national anthems
◆ Berlin Wall
◆ chancellor
◆ unification
◆ currency
◆ Holocaust

People and Places

◆ Adolf Hitler
◆ Poland
◆ East Germany
◆ West Germany
◆ Berlin
◆ East Berlin
◆ West Berlin
◆ Helmut Kohl
◆ Hamburg
◆ Munich
◆ Dachau
◆ Katarina Witt

Imagine what it would be like if you were not allowed to visit family or friends who lived nearby. How would you feel if it were not possible to call them on the telephone? This is what it was like in Germany before 1989. At that time Germany was divided into two countries. It had been divided soon after World War II ended. Today Germany is one country again. In this chapter you will learn why Germany was divided. You will learn how a united Germany is now an industrial leader.

Why Were There Two Germanys?

In 1933 Adolf Hitler became dictator of Germany. Hitler made all laws for the country. There was little freedom in Hitler's Germany. Hitler made Germany's army larger and stronger. In 1939 Germany attacked its neighbor, Poland. This was the start of World War II. Millions of soldiers from many nations fought in World War II. During the war large areas of Europe

GERMANY

Germany has several large rivers that are used for trade and transportation. Which two cities in Germany are located on rivers?

Germany's flag

were destroyed. Finally in 1945, Germany was **defeated**. The war ended.

The United States, Great Britain, France, and the Soviet Union had defeated Germany in 1945. After the war those four nations controlled different parts of Germany. The Soviet Union controlled the eastern part. It became East Germany. It had a Communist government. There was little freedom in East Germany. The rest of Germany became the country called West Germany. The people in both countries spoke German. But the two Germanys had their own laws, **national anthems**, flags, and money. Look at the map on this page of the two Germanys.

The United States helped rebuild West Germany after the war. It became a strong industrial country. It also became a democracy. People in West Germany had a much higher standard of living than the people of East Germany. They also enjoyed more freedom.

Berlin had been the capital of Germany before 1945. Berlin was in East Germany. After the war, Berlin was divided into two cities. East Berlin was a Communist city. It was East Germany's capital. West Berlin was a

Divided Germany and divided Berlin

The Berlin Wall was torn down in 1989. Also at that time people were allowed to travel between East Germany and West Germany.

German Chancellor Helmut Kohl

free city inside East Germany. The people of West Berlin enjoyed freedom and a high standard of living.

Millions of East Germans wanted more freedom and a higher standard of living. So they ran away to West Berlin. From West Berlin they could go to West Germany. The Communist leaders decided to stop people from leaving East Berlin. In 1961, they built the **Berlin Wall**. The Berlin Wall separated East and West Berlin. It was 99 miles long. The Berlin Wall made it very hard for East Germans to escape to West Berlin.

For many years the Communists kept the East Germans and the West Germans apart. Families could not visit each other. Friends could not make telephone calls to each other. But Germans dreamed about a time when their country would be united again.

How Did Germany Become a United Country?

In 1989 the East Germans tore down the Berlin Wall. People in the two Berlins could visit each other again. Berlin became one city again. On October 3, 1990, East Germany and West Germany became one united nation. People throughout Germany danced with joy. October 3 is now celebrated as a holiday each year. This holiday is called the Day of German Unity.

In December 1990, all Germans voted in national elections. Helmut Kohl became the **chancellor**, or leader, of Germany. He had been West Germany's chancellor for many years.

Most Germans are happy about **unification**, or uniting into one country. But unification has been difficult. West Germany's money became the **currency**, or money, for all of Germany. So East German money had to be changed. Products in the united Germany cost more than products had cost in East Germany. The people of East Germany have found it hard to buy these more expensive products. Many East Germans lost their jobs and could not find work in the united country.

The people of West Germany have had to pay higher taxes in order to pay for unification. These taxes have paid for new roads, railroads, and factories in East Germany.

Even though unification has not been easy, Germans are very proud of their united country. They are happy that Berlin is a united city. They are also happy that Berlin is the nation's capital again.

Tourists at the Brandenburg Gate in Berlin

Germany's Landforms, Resources, and Economy

The North Sea and the Baltic Sea are north of Germany. Germany has ports on these seas. The seas are important for trading and fishing. The seas also give most parts of the country a mild climate.

The North European Plain covers northern Germany. The central part of the country has hills and low mountains. The south has many mountains. The tall Alps cover part of southern Germany.

Germany has many rivers. The Rhine River and its many canals are the country's busiest waterways.

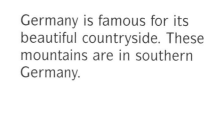

Germany is famous for its beautiful countryside. These mountains are in southern Germany.

This worker is testing computer parts in a factory in Germany.

Education is important in Germany. This classroom is in a high school in what was once East Berlin.

Large ships on the Rhine carry oil, coal, iron, steel, and other products to different parts of Germany.

Germany is not rich in natural resources. But it has many forests. It has some coal and a lot of iron. Germany uses its coal to make about one third of its electricity. It imports large amounts of oil. It also imports many kinds of raw materials for its factories.

Today Germany is the industrial leader of Europe. There are factories in many parts of the country. German factories make steel, cars, and planes. They make machines, computers, tools, and many other products. Germans sell products to many countries.

Only a small part of Germany's population works at farming. The climate is too cool for farmers to grow enough food. Germany imports one third of its food.

Germany's People, Cities, and Future

Germany has more than 83 million people. It has more people than any other country in Western Europe. The Germans are hard workers who feel proud when they do good work. But they also try to enjoy life. They enjoy good food. They also like to sing and dance with friends. Education is important to Germans. All children spend at least nine years in school.

Most Germans live in cities. Most cities have excellent subways and other transportation. Germans are very proud of Berlin. It has the largest department store in all of Europe. Berlin also has a special museum about the Berlin Wall. Hamburg is a large northern port. Munich, in the south, has many theaters and museums.

Dachau is a city near Munich. Many Jews were killed there during World War II. During the war the Germans tried to kill all the Jews of Europe. The killing of Jews and other groups of people during the war is called **Holocaust**. By the end of the war, six million Jews had been killed. Dachau has a museum that teaches about the Holocaust.

Germany is a rich country but it must solve some big problems. Many immigrants have moved to Germany. There are not enough jobs for them.

Unemployment is a problem in Germany today. The country does not have enough houses for all of its people. Pollution is another big problem. Factories have caused air pollution. River water has also become very dirty from all the factories. Germans are working to solve these problems. They want Germany to continue to be a great industrial country.

A highway in Germany

Chapter Main Ideas

1. Germany was divided into two countries after World War II. In 1990 unification made Germany one country again.
2. Germany is an industrial leader. But it must import food and raw materials.
3. Pollution and unemployment are problems in Germany today.

BIOGRAPHY

Katarina Witt (Born 1965)

Katarina Witt grew up in East Germany. She started ice-skating when she was five years old. The East Germans believed young Katarina could be an Olympic winner. So Communist leaders paid the country's best coach to train her. In 1984 Witt skated in the Olympic Games. She won a gold medal for East Germany. Four years later she won a second gold medal at the Olympics. The Communists rewarded her with cars and apartments.

Witt made many Germans angry when she spoke out against unification. Yet her life changed after unification. Witt could now decide for herself when or where she would skate. She was able to earn much more money. In 1992 Witt skated in the Olympics again. She skated for a united Germany, but she did not win a medal. Witt continues to skate in ice shows in Germany, the United States, and in other countries.

Journal Writing
Write a paragraph in your journal about Katarina Witt. Tell how Katarina's life changed after the unification of Germany.

◆ Vocabulary

Find the Meaning   On your paper write the word or words that best complete each sentence.

1. When Germany was **defeated** in 1945, Germany had _____.

    won the war   lost the war   built new cities

2. A **national anthem** is a _____ for your country.

    song of praise   special food   dance

3. The **unification** of Germany means the country became _____.

    divided   more developed   united

4. The **chancellor** is the _____ of Germany.

    money   national anthem   leader

5. **Currency** is the _____ used by a country.

    language   money   culture

◆ Read and Remember

Write the Answer   Write one or more sentences to answer each question.

1. How was Germany divided after World War II?

2. How was the city of Berlin divided?

3. Why did many East Germans want to go to West Berlin?

4. Why did Communists in East Germany build the Berlin Wall?

5. When did Germany become a united country again?

6. Why was unification hard for the German people?

7. What is the main landform in southern Germany?

8. What are some products that are made in German factories?

9. What must Germany import?

10. What are two problems in Germany today?

11. What did Germany do in 1939 that started World War II?

12. Who was the dictator of Germany before and during World War II?

13. Which four nations controlled parts of Germany after World War II?

14. What did the United States do for Germany after World War II?

15. What is the busiest waterway in Germany today?

◆ Think and Apply

Sequencing   Number your paper from 1 to 5. Write the sentences to show the correct order.

Germans voted for Helmut Kohl to lead a united Germany.

Germany attacked Poland and World War II began.

The Berlin Wall was torn down.

After Germany was defeated in World War II, the Soviet Union started a Communist government in East Germany.

Communists built the Berlin Wall in 1961.

◆ Journal Writing

Imagine that you were a German who lived in East Berlin in 1989. Write a paragraph that tells what you would have done after the Berlin Wall was torn down.

A **political map** helps you learn about countries. The map key on a political map uses symbols to show borders, cities, rivers, and other places.

The political map on this page shows unified Germany and its neighbors. The map shows the location of Germany's capital, two other cities, and three major rivers.

Study the political map and its map key. Then answer each question.

1. What three countries are to the east of Germany?

2. What seas are north of Germany?

3. What city in Germany is in the east?

4. What city in Germany is in the south?

5. What two countries are to the south of Germany?

6. What city is in northern Germany?

7. What countries are west of Germany?

8. What three rivers are in Germany?

9. What is the capital of Germany?

10. Which of the countries shown on the map does not share a border with Germany?

CHAPTER 16

Italy and Mediterranean Europe

Think About As You Read

1. How are the four countries in the Mediterranean region alike?
2. How are the three regions of Italy different from each other?
3. Why is Vatican City important?

New Words

- mountainous
- favorable balance of trade
- arches
- ruins
- monuments
- pope

People and Places

- Greece
- Sicily
- Sardinia
- Apennine Mountains
- Po River
- Po Valley
- Rome
- Tiber River
- Colosseum
- Vatican City
- Bologna
- Florence
- Venice

Italy is one of the oldest countries in Europe, but it is also one of the newest. It is old because it began more than 2,000 years ago. Italy is new because it did not become a united country until 1870. Today Italy is one of several European countries on the Mediterranean Sea.

The Mediterranean Region

Italy, Spain, Portugal, and Greece are four West European nations in the Mediterranean region. Mountains separate them from the rest of Europe. About one third of Western Europe's people live in this region. Italy has the largest population in this region. Italy has more than 57 million people.

The four Mediterranean countries are alike in some ways. They all have a Mediterranean climate. So their winters are rainy and their summers are long, hot, and dry. Mountains and hills cover much of the land. They have few rivers that can be used for transportation. These countries have few natural resources.

MAP KEY
- City
- ⊛ Capital city
- – · – Border between countries
- ~ River

FRANCE

Milan
Po River
Bologna

ALPS

ITALY

APENNINE MOUNTAINS

Tiber River
Rome

Naples

Palermo

Vardar River

Thessaloníki

GREECE
PINDUS MTS.
Athens

Oporto
Bilbao
PYRENEES
Ebro River
Douro River
PORTUGAL
Madrid
Tagus River
Lisbon
SPAIN
Barcelona
Valencia
Guadalquivir R.
Seville

Mediterranean Sea

The four countries on this map have many things in common. They have a Mediterranean climate, mountainous land, and few natural resources. Which of these four countries does not have a coast on the Mediterranean Sea?

More people work at farming in the Mediterranean region than in the northern parts of Western Europe. Olive trees grow throughout this region. These countries export olives and olive oil. Except for Italy, this region has less industry than the rest of Western Europe. Spain, Portugal, and Greece have lower standards of living than the rest of Western Europe.

Religion is important to the people in this region. Most of the people in Spain, Portugal, and Italy are Roman Catholic. Most Greeks belong to the Greek Orthodox Church.

Olive trees

Italy: A Mediterranean Country

Italy is a long peninsula in the Mediterranean Sea. The country is shaped like a boot. The islands of Sicily and Sardinia in the Mediterranean Sea belong to Italy.

Most of Italy is **mountainous**. The Alps separate Italy from France and Switzerland. The Apennine Mountains cover a large part of the country from north to south. Some Apennines Mountains are volcanoes.

There are three regions in Italy. The northern region is the country's industrial region. Factories in the northern region make cars, steel, machines, shoes, and many other products. Some of the world's best sports cars are made in this region. Waterpower in this region is used to make about one fourth of Italy's electricity. The Po River is in northern Italy. Plains around the river form the Po Valley. The Po Valley has Italy's best farmland. Northern Italy is the country's richest region.

The second region is central Italy. This region surrounds the city of Rome. Rome is Italy's capital and its largest city. Plains and low hills near the western coast have good farmland.

Southern Italy is the third region. It is the poorest part of Italy. It is mountainous. The soil is not good for farming. So it is difficult to grow food in this region. People raise sheep and goats. Some people produce olive oil from olive trees. This region has few factories. Many people have moved from southern Italy to the north to get factory jobs. In the early 1900s, many people from southern Italy moved to the United States.

Today about ten percent of Italy's people earn a living by farming. These farmers grow most of the food Italy needs. Farmers grow large amounts of grapes for wine. Italy makes more wine than any other country.

About one third of the population works at factory jobs. More than half of the people earn a living at

Italy

Italy's flag

Northern Italy has some of the country's best farmland. The Alps are also found in this part of Italy.

Rome is a mixture of the very old and the very new. The ruins of the Colosseum are next to modern apartment buildings.

The Colosseum

service jobs. Many of these people work in the tourism industry. Italy earns billions of dollars each year from tourism.

Italy earns a lot of money from trade. Some of the products Italy exports are cars, shoes, cheese, wine, and olive oil. Italy has a **favorable balance of trade**. This means it exports more than it imports.

Italy's History and Government

About 2,500 years ago, people began to build the city of Rome. The city was built on seven hills near the Tiber River. The Romans then conquered much land and many people to create a huge empire. The Romans were great builders. They built many temples, theaters, roads, and **arches** in Rome and throughout their empire. The Romans ruled their empire for 900 years. In the year 476, the Roman Empire fell apart.

After the Roman Empire fell apart, Italy was made of many small states. Each state had its own laws and its own ruler. In the 1800s people began working to unite all the states. In 1870 Italy became one nation.

Today Italy is a democracy. A prime minister leads the country. People vote for members of Parliament.

Rome, Vatican City, and Other Cities in Italy

Two thirds of Italy's people live in cities. Rome has about 3 million people. It has fewer factories than other large cities in Europe. Most Romans work at government and service jobs. Others work in hotels and restaurants for tourists.

Rome is both a modern city and a very old one. The city has modern buildings, lots of cars, and a subway system. But in Rome you can also see **ruins**, or very old buildings, from the time of the Roman Empire. There are very old temples, fountains, statues, and **monuments**. You can see the Colosseum that the Romans built long ago. This huge theater had enough seats for 50,000 people. Today visitors can drive around the Colosseum on a modern six-lane highway.

Ancient Romans built arches to honor their leaders.

Inside the city of Rome is a tiny independent country. That country is Vatican City. It is also called the Vatican. It has its own flag, laws, money, and stamps. Fewer than 1,000 people live in this tiny country. But it is important because it has the government for the Roman Catholic Church. There are almost one billion people who belong to the Roman Catholic Church today. The **pope**, the world leader of the Catholic Church, is the ruler of the Vatican. Millions of people visit the Vatican each year.

Italy has other important cities. Bologna has Europe's oldest university. Florence, in central Italy, is famous for its paintings and statues. The city of Venice is in northern Italy. Venice was built on about 120 islands. People travel through the many canals of Venice on boats. The people of Italy are proud of their cities, their long history, and their modern industries.

Pope John Paul II in Vatican City

Chapter Main Ideas

1. Italy, Spain, Portugal, and Greece are the four Mediterranean countries of Western Europe. Italy has the most people and the highest standard of living.
2. Italy's history began in Rome 2,500 years ago.
3. Northern Italy is a rich industrial region. Southern Italy is a poorer region where people work at agriculture.

◆ Vocabulary

Match Up   Finish the sentences in Group A with words from Group B. On your paper write the letter of each correct answer.

Group A

1. A country that has many mountains is _____.

2. Italy exports more goods than it imports, so it has a _____.

3. The leader of the Roman Catholic Church is the _____.

4. A building that was built to remember an event is a _____.

5. In Rome you can see very old buildings, or _____, from long ago.

Group B

A. favorable balance of trade

B. ruins

C. mountainous

D. monument

E. pope

◆ Read and Remember

Where Am I?   Read each sentence. Then look at the words in dark print for the name of the place for each sentence. On your paper write the name of the correct place.

| Rome | Vatican City | Greece | Bologna | Venice | Sicily |

1. "I am in a country where most people follow the Greek Orthodox religion."

2. "I am in a city in northern Italy where people travel on boats through canals."

3. "I am in Italy's capital and largest city."

4. "I am in an Italian city with the oldest university in Europe."

5. "I am in a tiny independent country inside the city of Rome."

6. "I am on an island that is part of Italy and is in the Mediterranean Sea."

Finish Up   Choose the word or words in dark print that best complete each sentence. Write the word or words on your paper.

Tiber River   Colosseum   industry   Apennines   Sardinia

1. The Alps and the _____ are two mountain chains in Italy.

2. Northern Italy has more _____ than southern Italy.

3. The islands of Sicily and _____ belong to Italy.

4. The city of Rome is on the _____.

5. Long ago the Romans built a huge theater called the _____.

◆ Think and Apply

Compare and Contrast   Copy the Venn diagram shown below on your paper. Then read each phrase below. Decide whether it tells about northern Italy, southern Italy, or all of Italy. Write the number of the phrase in the correct part of the Venn diagram on your paper.

1. industrial region

2. Mediterranean climate

3. most people work at agriculture

4. richest region

5. poorest region

6. about 57 million people

7. most people are Roman Catholics

8. poor soil for farming

9. fertile Po Valley

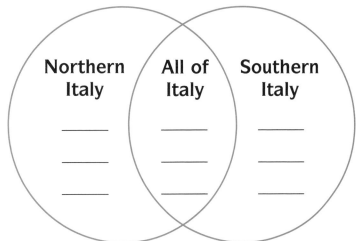

◆ Journal Writing

Write a paragraph in your journal that tells about the countries of the Mediterranean region. Tell one fact about each country in the region.

Italy and most other countries have several different landforms. Some places have low plains. Other areas have tall mountains. A **physical map** shows how high the land is in different places. From a physical map you can learn where there are low plains, hills, and high mountains. An **elevation key** is part of a physical map. The key uses different colors to show different elevations.

On your paper write the answer to each question.

1. What is the elevation of the Alps near France?

0–650 feet   650–1,500 feet   6,500–13,000 feet

2. What is the elevation of Mt. Etna?

less than 6,500 feet   10,902 feet   more than 13,000 feet

3. What is the highest elevation of the Apennine Mountains?

0–650 feet   650–1,500 feet   more than 1,500 feet

4. What is the elevation of Venice?

less than 650 feet

more than 1,500 feet

more than 6,500 feet

5. What is the elevation of Rome?

0–650 feet

650–1,500 feet

more than 13,000 feet

6. What can we conclude about the elevation of Italy's cities?

their elevation is high

their elevation is low

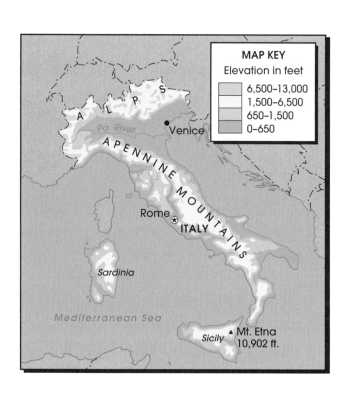

CHAPTER 17

Understanding Western Europe

Where Can You Find?
Where can you find the leaders of a united Western Europe?

Think About As You Read

1. How has NATO helped Western Europe?
2. How are the countries of Western Europe solving energy and pollution problems?
3. How is the European Union helping Western Europe become more united?

New Words

- ◆ passport
- ◆ NATO
- ◆ Partnership for Peace
- ◆ hydroelectric power
- ◆ solar energy
- ◆ geothermal energy
- ◆ recycle
- ◆ European Community
- ◆ European Union
- ◆ euro

People and Places

- ◆ Turkey
- ◆ Norway
- ◆ Belgium
- ◆ Sweden
- ◆ Iceland
- ◆ Brussels

Imagine taking a train ride through several countries in Western Europe. At most borders, guards would not check your **passport**. Traveling from country to country in Western Europe is now almost as easy as traveling through different states of the United States. In the past it was not so easy. Today the countries of Western Europe are working together in new ways.

Western Europe Today and the Need for NATO

Today there is peace in Western Europe. But for hundreds of years, there were wars in this region. Europeans fought wars to rule their continent. They also fought wars to control colonies in Africa, Asia, and the Americas. In the 1900s most countries in Europe fought in World War I and World War II.

After World War II, the countries of Western Europe were afraid that the Soviet Union and other Communist countries would attack them. So they formed an organization called **NATO** to protect

NATO AND THE PARTNERSHIP FOR PEACE

MAP KEY
◻ NATO member
◻ Partnership for Peace member

There are 16 countries in NATO. The United States and Canada are two members of NATO not shown on this map. Partnership for Peace (PfP) countries have joined with NATO to bring peace to countries that have wars. What are the names of two PfP nations?

NATO's flag

Air force planes from Turkey and the United States fly together for NATO.

themselves. NATO stands for the North Atlantic Treaty Organization. Thirteen Western European nations belong to NATO. Canada, Turkey, and the United States are also members. Soldiers from NATO nations will fight together if an enemy attacks a member.

Today communism is no longer a problem for Western Europe. So NATO's new goal is to bring peace to countries that have wars. Many countries in Eastern Europe are now working with NATO. They are not yet NATO members. They are part of a program called the **Partnership for Peace**, or PfP. Soldiers from PfP countries work with NATO soldiers. Some countries such as Poland and Hungary may soon be allowed to join NATO. Russia will be allowed to work more closely with NATO in the future.

Energy and Pollution Problems

Western Europe is an important industrial region. But many of the natural resources in this region have been used up. Today most countries in Europe import large amounts of raw materials from far-off nations.

Western Europe needs lots of energy for its cars, homes, and factories. Norway and the United Kingdom

138

get oil from the North Sea. But many other countries must import oil. Many European countries do not want to import oil from other places because it is expensive. So they are trying to find other ways to make energy. Today France, Belgium, Switzerland, and other countries are using nuclear energy to make electricity. Italy, Sweden, Switzerland, and some other countries are using **hydroelectric power**, or electricity made from waterpower. Some countries like Greece are using energy from the sun, or **solar energy**. A few countries like Iceland have **geothermal energy**, or energy from hot places inside the earth.

Europeans want to end pollution.

Pollution is a very big problem in Western Europe. Cars and factories are causing air pollution. Air pollution becomes part of rain and snow and causes acid rain. Acid rain is destroying forests, fish, and lakes in many European countries.

Today Western Europe is trying to end pollution. Many countries have passed clean air laws. Cars and factories must be built so they send less pollution into the air. Some nations have passed clean water laws. The United Kingdom was one of those nations. The British worked hard to make the dirty Thames River clean again. Some beaches on the Mediterranean Sea have also been made clean and safe.

Europeans are working to protect their environment. Most people **recycle** old cans, bottles, paper, and plastic. Factories are using less paper, cardboard, and

This place in southern Iceland uses geothermal energy to make electricity.

The European Union

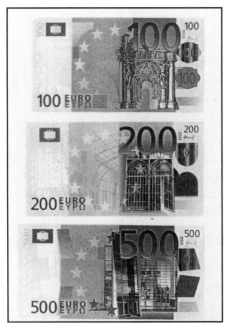

Europe's new currency, the euro

plastic to wrap products such as candy, toothpaste, and televisions. By using less wrapping materials, there will be less trash to harm the environment.

Working for a United Europe

Some people hope that one day all the countries of Europe will unite and become one country. In 1957 six countries of Western Europe began working together as the **European Community**. Members of the European Community worked to have more trade with each other. More nations later joined the European Community. In 1994 the European Community became the **European Union**. It is also called the EU. Fifteen Western European countries are part of the EU. The leaders of the EU meet in Brussels, Belgium.

The European Union is working for a united Europe. People no longer need passports to cross the border from one member nation to another. People from one EU nation can live or work in any other EU country.

The European Union helps trade between its members. There are no tariffs on goods that are bought or sold between EU members. The European Union now exports more goods than the United States.

Today every country in Western Europe has its own money. In 1999 all EU members will use the same currency. The new currency will be called the **euro**.

The people in Western Europe are proud of their countries. They are proud of their cultures. But they are also proud to be Europeans. Can the EU join all the countries together to form one country? Many people are working toward this goal. Perhaps one day there will be a United States of Europe.

Chapter Main Ideas

1. Western Europe is a developed region. It depends on imported raw materials for its factories.
2. Many countries are working to solve the problems of air pollution, acid rain, and water pollution.
3. The countries of Western Europe are working together through NATO and the European Union.

◆ Vocabulary

Forming Word Groups   Copy the chart shown below on your paper. Read each heading on the chart. Then read each word in the vocabulary list on the left. Form groups of words by writing each vocabulary word under the correct heading on your paper.

Vocabulary List	Europeans Working Together	Types of Energy
NATO	1. _____	1. _____
hydroelectric		
euro	2. _____	2. _____
solar		
Partnership for Peace	3. _____	3. _____
European Community	4. _____	4. _____
geothermal		
European Union	5. _____	
nuclear		

◆ Read and Remember

Complete the Chart   Copy the chart shown below on your paper. Use facts from the chapter to complete the chart. You can read the chapter again to find facts you do not remember.

Problems in Western Europe

	Europe Needs Raw Materials	Europe Needs a Lot of Energy	Pollution and Acid Rain
What is the problem?			
How are people solving the problem?			

Write the Answer   Write one or more sentences to answer each question.

1. Why was the North Atlantic Treaty Organization, NATO, formed?

2. What is NATO's new goal?

3. How are former Communist nations working with NATO?

4. How does the European Union help trade between its members?

5. What three members of NATO are not Western European countries?

6. What kinds of energy are the countries of Western Europe using to make electricity?

◆ Think and Apply

Finding Relevant Information   Imagine you are telling your friend how the nations of Western Europe are working for a united Europe. Read each sentence below. Decide which sentences are relevant to what you will say. On your paper write the relevant sentences you find. You should find three relevant sentences.

1. In 1957 six nations in Western Europe began working together for better trade.

2. People who live in a European Union nation can live or work in any other European Union country.

3. Today communism is no longer a problem for the countries of Western Europe.

4. In 1999 all European Union countries will have the same currency.

5. The British passed clean water laws to end pollution in the Thames River.

6. Today there is peace in Western Europe.

7. Western Europe is an important industrial region.

◆ Journal Writing

Journal Writing   Write a paragraph about two problems in Western Europe today. Tell how people are working to solve those problems.

Eastern Europe and Russia and Its Neighbors

Budapest, Hungary

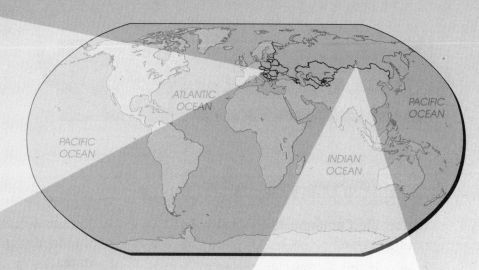

Lake Baikal, Russia

DID YOU KNOW?

▲ The world's longest train ride is 6,000 miles on Russia's Trans-Siberian Railroad.

▲ Moscow has the world's busiest subway system. It is used about 4 billion times each year.

▲ Lake Baikal in Russia is the world's deepest lake. More than 350 rivers flow into the lake, but only one flows out.

▲ Budapest, Hungary, is really two cities. They are Buda and Pest. They are on different sides of the Danube River.

WRITE A TRAVELOGUE

Imagine you are going to travel by train through Eastern Europe and Russia and its neighboring countries. Keep a travelogue about your trip. Which countries would you want to visit? Before reading Unit 4, write a paragraph about why you might want to visit those countries. After reading the unit, write two paragraphs about the ways that people and goods can move through this region.

THEME: MOVEMENT

CHAPTER 18

A Changing Region

Think About As You Read

1. How did Communists win control of this region?
2. Why is this a changing region?
3. What are the landforms, climates, and resources of this region?

New Words

- Commonwealth of Independent States
- czars
- Russian Revolution
- command economy
- free market economy
- continental climate

People and Places

- Russia
- Estonia
- Latvia
- Lithuania
- Ural Mountains
- Danube River
- Black Sea

Eastern Europe and Russia and its neighboring countries form the region that is east of Western Europe. It is a region that is changing in the way people live, work, and choose government leaders.

A Huge Region

Eastern Europe and Russia and its neighbors form a huge region with land on two continents. The region covers the eastern part of Europe and the northern part of Asia. This region can be divided into three groups of countries. The first group is the countries of Eastern Europe. The second group includes three countries along the Baltic Sea—Estonia, Latvia, and Lithuania. These Baltic countries are in northern Europe. Russia and 11 smaller countries are the third part of the region. These 12 countries have formed a group called the **Commonwealth of Independent States**, or C.I.S.

This region contains many countries in both Europe and Asia. What mountain chain divides the western part of Russia from Siberia?

For more than forty years, Communists controlled the entire region. Today these countries do not have Communist governments. How did this happen?

The History of the Region

For hundreds of years, Russia was a country that was ruled by kings. The kings were called **czars**. They made all laws. Most Russians were very poor. The people did not have enough food. The czars did little to help the poor. By 1917 many Russians were angry. They wanted a different government. So in that year, angry Russians started a war called the **Russian Revolution**. During this fight Communists won control of Russia.

The Russians later took control of their neighboring countries. They formed one country called the Soviet Union. A Communist government ruled the entire Soviet Union. During World War II, the Soviet Union

A Russian czar and his family

The Russian people celebrated the end of communism in their country in 1991.

The three Baltic nations are the only part of the former Soviet Union that did not join the C.I.S.

MAP KEY
C.I.S.
Former Soviet Union

took control of the three Baltic countries. They were forced to become part of the Soviet Union.

After the war, the Soviet Union forced the nations of Eastern Europe to have Communist governments. These countries were not part of the Soviet Union. But the Soviet Union controlled their governments.

There was little freedom in Eastern Europe and the Soviet Union. The Communists did not allow free elections. People could not speak out against their government. People could not travel to other countries. Freedom of religion was not allowed. People were afraid to practice their religions.

In these Communist countries, the government owned all farms, factories, and businesses. The government decided what crops farmers should grow. The government told factories what to make and what prices to charge. The government decided how much the workers should be paid. This type of controlled economy is called a **command economy**.

By 1989 many people wanted to change their Communist governments. Communism first ended in Poland. Then the Berlin Wall opened in East Germany. Other countries in Eastern Europe formed new governments that were not controlled by Communists. At the end of 1991, communism ended in the Soviet Union. The huge country fell apart. Russia became an independent country again. The other 14 countries that had been part of the Soviet Union also became free. Eleven of them formed the C.I.S. with Russia. The three Baltic countries did not join the C.I.S. Many countries in the region are trying to have democratic governments.

How Are Countries in the Region Alike and Different?

The countries of this region are different from each other. They have their own governments and their own languages. But they are also alike in some ways. All of these countries are less developed than the nations of Western Europe. People in Eastern Europe have a lower standard of living than the people in Western Europe. Most of the people living in the C.I.S. have an even lower standard of living. Eastern Europe has more industry and better farming than the C.I.S. But in many places throughout this region, people still use animals, not machines, to do farm work.

A free market economy helps businesses such as this shipyard in Poland.

The countries in this region are alike in another way. Today Communists no longer control the region. Most of the countries are changing to a **free market economy**. In a free market economy, people own and control companies, factories, farms, and businesses. The government does not own or control them. The United States, Canada, and Western Europe have free market economies. Many countries in the region are finding it difficult to change to a free market economy. Many products have become more expensive. Many people do not have jobs.

Landforms, Climates, and Resources

The North European Plain covers the northern part of this region. There is a lot of good farmland on the plains of Europe. The region also has mountains. The low Ural Mountains separate Europe from Asia. Most mountains are in the southern part of the region. Look at the map on page 145. Which areas are covered with plains? Which areas have mountains?

The Danube River

The Danube River is the most important river in this region. The Danube is 1,776 miles long. It flows south from Germany through many countries in Eastern

Parts of Eastern Europe are mountainous. People enjoy visiting the mountains during the winter to go skiing.

People in this region can speak out against their leaders now that communism has ended.

Europe. Then it runs into the Black Sea. The Danube is used for trade and transportation. There are many ports on this river. Like the Rhine River, water pollution is a big problem on the Danube.

Most of this region has a cold climate. Warm ocean currents from the Atlantic Ocean do not warm the region because the ocean is too far away. Many places have a **continental climate**. This means winters are long and very cold. Summers are short and hot. As you move north and east, the climate gets much colder. This region also gets less rain than Western Europe. In the southern part of the region, the climate is warmer. Some southern areas have a Mediterranean climate.

Parts of this region are rich in natural resources. Poland and some other Eastern European countries have coal. Russia has coal, oil, diamonds, and many other resources.

The end of communism has changed the way people live and work in Eastern Europe and Russia and its neighbors.

Chapter Main Ideas

1. Eastern Europe and Russia and its neighbors cover parts of Europe and Asia.
2. Communists no longer control this region. Countries are starting free market economies.
3. Plains cover much of the region. The climate is colder than the climate in Western Europe.

◆ Vocabulary

Finish Up   Choose the word or words in dark print that best complete each sentence. Write the word or words on your paper.

free market economy   **command economy**   **continental climate**
Commonwealth of Independent States   **czars**

1. For hundreds of years, kings called _____ ruled Russia.

2. A _____ means there are long, cold winters and short, hot summers.

3. Russia and 11 countries that were part of the Soviet Union are now part of the group called the _____.

4. In a _____, the government decides what factories will make and what workers will earn.

5. In a _____, people own factories, farms, and businesses.

◆ Read and Remember

Write the Answer   Write one or more sentences to answer each question.

1. How did Eastern Europe become a Communist area?

2. How did the Communists not allow people freedom?

3. How did communism end in this region?

4. What are the main landforms of the region?

5. How does the Danube River help Eastern Europe?

6. What kinds of climates does the region have?

7. What are the three groups of countries that make up this region?

8. What are some of the natural resources of this region?

9. Who were the czars?

10. What landform separates Europe from Asia?

11. How are all the countries of this region alike?

12. What is a big problem with the Danube River?

13. Why did the Russian people start the Russian Revolution?

14. What three countries were forced to become part of the Soviet Union after World War II?

◆ Think and Apply

Categories   Read the words in each group. Decide how they are alike. Find the best title for each group from the words in dark print. Write the title on your paper.

End of Communism   **Free Market Economy**   **Danube River**
Command Economy   **Natural Resources**   **Baltic Countries**

1. government decides what prices to charge
   government decides what salaries to pay
   government decides what factories will produce

2. people own their own businesses
   people decide what factories will produce
   people decide prices to charge and salaries to be paid

3. Poland changed its government
   the Berlin Wall was torn down
   the Soviet Union broke apart

4. many ports
   begins in Germany
   flows into the Black Sea

5. coal
   iron
   oil

6. Estonia
   Latvia
   Lithuania

◆ Journal Writing

Write a paragraph in your journal that explains how a command economy is different from a free market economy.

CHAPTER 19

Poland: A Changing Industrial Nation

Where Can You Find?
Where can you find some of the oldest salt mines in the world?

Think About As You Read

1. How have Poland's plains helped and hurt the country?
2. How did Solidarity help Poland?
3. Why has it been hard for Poland to have a free market economy?

New Words

- rye
- consumer goods
- concentration camps
- trade unions
- Solidarity
- legal
- privatization
- strike

People and Places

- Czech Republic
- Slovakia
- Ukraine
- Vistula River
- Warsaw
- Gdańsk
- Wieliczka
- Pope John Paul II
- Lech Walesa

Poland is the largest country in Eastern Europe. It is an industrial country. But Poland is different from Western Europe because it is difficult for people to own cars. As you read, find out why fewer than 7 million people own cars in Poland.

Poland's Geography and Resources

Most of Poland is covered with plains and rolling hills. In fact the name Poland means "plains." These plains have both helped and hurt Poland. They help Poland by giving the country good farmland. But the flat plains have made it easy for other countries to attack Poland from the east and from the west.

The Baltic Sea is north of Poland. There are many beaches and small lakes near the sea. Mountains are to the south. They separate Poland from three of its neighbors, the Czech Republic, Slovakia, and Ukraine. Look at the map on page 152. Name four other countries that share borders with Poland.

The North European Plain is the largest landform in Poland. What mountains are located in southern Poland?

POLAND

Poland's flag

A farm in Poland

The Vistula River is the most important river in Poland. It runs from the mountains in the south to the Baltic Sea. This river is used for transportation. Warsaw, Poland's capital and largest city, is on the Vistula River. The port city of Gdańsk is on the Baltic Sea near this river.

Coal is Poland's most important resource. There are rich coal fields in the southern mountains. Poland also has copper, silver, salt, and other resources. Salt mines in Wieliczka are among the oldest in the world. Poland must import two important resources, oil and iron.

People, Culture, and Economy

Almost 39 million people live in Poland. They are called Poles. They speak the Polish language. Very few immigrants from other countries live in Poland.

Religion and education are very important to Poles. Most Poles are Roman Catholics. Pope John Paul II, the world leader of the Catholic Church, grew up in Poland. All children in Poland must go to school for at least nine years. Almost everyone knows how to read and write.

In the past most Poles were farmers. Today only about one fourth of the people work at farming. Potatoes and **rye** are the most important crops. Rye is a grain that is used to make dark bread. These crops grow well in Poland's cool climate.

After World War II, Communist leaders turned Poland into an industrial country. Many factories were built. Now one third of the people work at factory jobs. Polish factories make products such as steel, cement, and machines. But they do not make many **consumer goods**. Consumer goods are things people buy for their own needs. Clothing and cars are two kinds of consumer goods. Many Poles also work at service jobs.

Apartments in Poland

Poland has a lower standard of living than Western Europe. Poles earn low salaries, and consumer goods are expensive. Poles must spend most of their money on food. Fewer than 7 million people own cars. Most city people live in small apartments.

Poland's History

Poland has been conquered many times. In 1795 Germany, Austria, and Russia conquered Poland. The three countries divided Poland. In that year Poland disappeared from the map of Europe. Poland did not become a free nation again until after World War I.

During World War II, Germany and Russia conquered and divided Poland again. The German Army built six **concentration camps**, or death camps, in Poland. Most of Poland's 3 million Jews were killed in those concentration camps during the Holocaust. Three million Christian Poles also died during the war.

During World War II, this was a concentration camp in Poland.

After the war, Russia forced Poland to have a Communist government. Most Poles were not happy with their Communist government. Communist leaders made it hard for people to practice the Catholic religion. The Poles wanted the freedom to speak out against their leaders. They also wanted a free market economy.

In 1980 Polish workers began to form **trade unions**. Trade unions are groups that help workers. Sometimes they help workers get better salaries. All of the new Polish trade unions became part of a group called **Solidarity**. The Communists made it against the law to join Solidarity. Solidarity leaders were sent to jail. But millions of Poles continued to join Solidarity.

Members of Solidarity called for changes in the government and economy of Poland.

Lech Walesa and other Poles can now vote in free elections.

They hoped the unions would bring more freedom to Poland. Lech Walesa became the leader of Solidarity.

In 1989 the government said it was **legal** for people to join Solidarity. That same year Poles voted in the country's first free election. The Communists lost the election. Lech Walesa was elected president.

Poland began to change to a free market economy. The government began the **privatization** of factories. This means the government started selling the factories it owned to private companies.

People have much more freedom in Poland today. But changing to a free market economy has been very difficult. About 6 million people do not have jobs. Food and consumer goods are very expensive.

In 1995 Poles voted in free elections again. This time many Communists were elected to government positions.

The Future of Poland

Today Polish factories make more products than ever before. Poland exports more goods each year. But Poland still has a much lower standard of living than Western Europe. Most factories do not have modern technology. Most farmers do not use modern methods. Poland is trying to buy new technology from Western Europe and the United States. This will help Poland become a more modern country. Many American companies have started businesses in Poland.

Poland wants more friendship with Western Europe. So it is planning to become a member of NATO and the European Union.

Most Poles are unhappy that they still do not have enough consumer goods. They do not like the high prices they must now pay for food and goods. Will Poland become a Communist nation again? Will the country change back to a command economy? The people of Poland still have much work to do.

Chapter Main Ideas

1. Poland's plains have good farmland.
2. Communists lost control of Poland in 1989 when Solidarity leaders were elected in free elections.
3. Poland is trying to have a free market economy. But Poland needs modern technology to improve the standard of living of its people.

Poland now has a free market economy.

BIOGRAPHY

Lech Walesa (Born 1943)

Lech Walesa helped end communism in Poland. He saw how unfair Poland's Communist government was to Polish workers. So he told workers to join trade unions. He helped the unions unite to form Solidarity. Walesa became the leader of Solidarity.

In 1980 Walesa helped the workers win a **strike** against the government. But the government decided that Solidarity was illegal. Walesa was sent to jail. While in jail he won the Nobel Peace Prize for his work for Solidarity.

After Walesa was freed, he continued working to end communism in Poland. In 1989, Poland had its first free elections. Walesa was elected president. Walesa grew less popular when goods became very expensive. Walesa ran for president again and lost. He continues to speak out against communism in Poland.

Journal Writing
Write a paragraph in your journal that tells how Lech Walesa helped end communism in Poland.

◆ Vocabulary

Match Up Finish the sentences in Group A with words from Group B. On your paper write the letter of each correct answer.

Group A

1. A grain called _____ is used for making dark bread.

2. Goods that people buy for their own use are _____.

3. Death camps that were built by the German Army during World War II were called _____.

4. Groups that help workers get better salaries are _____.

5. Something that is allowed by law is _____.

6. When the government sells the factories it owns to private companies, there is _____.

Group B

A. trade unions

B. rye

C. legal

D. concentration camps

E. consumer goods

F. privatization

◆ Read and Remember

Finish the Paragraph Number your paper from 1 to 7. Use the words in dark print to finish the paragraph below. On your paper write the words you choose.

| disappeared | Lech Walesa | consumer | coal |
| Warsaw | plains | Catholic | |

Much of Poland is covered with __1__. The capital of Poland, __2__, is on the Vistula River. Poland's most important resource is __3__. Most Poles practice the __4__ religion. Poland is an industrial country, but it manufactures fewer

__5__ goods than Western Europe. Fewer than 7 million Poles own cars. In 1795,

Austria, Russia, and Germany took control of Poland. Poland __6__ from the map

of Europe until after World War I. Solidarity was formed in 1980. It is an organization

of trade unions. __7__ worked to end communism in Poland while he was the

leader of Solidarity.

◆ Think and Apply

Cause and Effect   Number your paper from 1 to 6. Write sentences on your paper by matching each cause on the left with an effect on the right.

Cause

1. Poland has flat plains, so _____.

2. Three million Jews were killed in the Holocaust, so today _____.

3. Poland has a cool climate, so _____.

4. After World War II, Poland became an industrial country, so _____.

5. In 1980 Polish workers wanted better salaries and more freedom, so _____.

6. Today Poland wants to have a free market economy, so _____.

Effect

A. farmers grow rye and potatoes

B. there is privatization of factories

C. it has been easy for enemies to attack Poland

D. there are about 10,000 Jews in Poland

E. they joined Solidarity

F. fewer people are farmers

◆ Journal Writing

Poland held free election in 1989. The Communists lost this election. Write a paragraph in your journal that tells how communism ended in Poland.

Understanding Line Graphs

Line graphs are used to show **trends.** Trends are changes that take place over a period of time.

The line graph on this page shows how Poland's imports and exports have changed between 1982 and 1994. Since communism ended in Poland in 1989, the country has had more trade with Western Europe. It has more imports and exports.

Look at the line graph on this page. Number your paper from 1 to 5. Then finish each sentence in Group A with an answer from Group B. Write the letter of the correct answer on your paper.

Group A

1. In 1982 Poland had _____ in exports

2. In _____ Poland's imports were $5.4 billion.

3. In 1990 Poland earned _____ from its exports.

4. Poland exported more than it imported every year except _____.

5. In 1994 Poland did not have a _____ balance of trade.

Group B

A. 1994

B. $13.5 billion

C. favorable

D. 1986

E. $4.96 billion

Poland's Imports and Exports

CHAPTER 20

Russia: The World's Largest Country

Think About As You Read

1. What landforms and climates are in Russia?
2. Why is transportation poor in many parts of Russia?
3. What problems does Russia have today?

New Words

- tundra
- steppes
- Russian Orthodox Church
- republics
- plots
- heavy industries
- glasnost
- shortage
- social problems
- alcohol abuse

People and Places

- Trans-Siberian Railroad
- Moscow
- Siberia
- Volga River
- Kremlin
- St. Petersburg
- Petrograd
- Leningrad
- Mikhail Gorbachev
- Boris Yeltsin
- Chechnya
- Lake Baikal

Imagine traveling across Russia by train. You would board the Trans-Siberian Railroad in Asia near the Pacific Ocean. For six days you would travel west through Russia. After traveling more than 5,000 miles, you would finally reach Moscow, the capital of Russia. You would still have to travel another 1,000 miles to cross the border of the world's largest country.

Russia's Landforms, Climates, and Resources

Russia is almost twice the size of the United States. Russia is on two continents. The western part is in Europe. The eastern part is in Asia and is called Siberia. The Ural Mountains separate Europe and Asia.

Most of Russia has very long, cold winters. It is cold because it is so far north. Snow covers much of the country for at least six months each year. The temperature in January is often below 0°F. As you move south, the climate becomes warmer. The region near the Black Sea has a mild Mediterranean climate.

The Trans-Siberian Railroad crosses Russia, the largest country in the world. Which two cities are connected by the railroad?

Russia's flag

RUSSIA

MAP KEY
- City
- ⊛ Capital city
- –·– Border between countries
- ∼ River
- ++++ Trans-Siberian Railroad

A village in Siberia

Most of Russia is covered with plains. Plains cover European Russia west of the Urals. Plains also cover a large part of Siberia east of the Urals. The plains can be divided into three regions. The northern plains are cold and icy. This icy land is called **tundra**. The tundra is covered with permafrost like northern Canada, which you read about in Chapter 3. Only small plants can grow on permafrost in summer. The second region is south of the tundra. Thick forests cover the land. South of the forests are grassy plains called **steppes**. This region has the best farmland in Russia.

Russia has some of the world's longest rivers. At 2,300 miles, the Volga River is the longest river in Europe. In the summer, rivers in Russia are used for shipping and transportation. But most rivers freeze in the winter. Trucks then use the frozen rivers as roads.

Russia has a long northern seacoast. But the water near the coast is filled with ice for many months during the winter. There are few ports that are good for shipping throughout the year. This makes it hard for Russia to trade with other countries.

Russia has more natural resources than any other country in the world. It has coal, oil, and many kinds

St. Petersburg is a port on the Baltic Sea. The city was built on more than 100 islands. Like Venice, this city has many canals.

of metals. But many resources are not being used. Most resources are in Siberia. The cold climate makes them difficult to mine. Also because of the cold, northern Siberia does not have good transportation for moving resources to factories.

Russia's People and Cities

Russia has 148 million people. This is a much smaller population than in the United States. Most Russians live in the European part of Russia. About three fourths of Russia's people live in cities. Russia's largest cities are in the European part of Russia.

Moscow is Russia's capital and its largest city. It has about 10 million people. Moscow is the center of the government and the economy. Russia's leaders meet in the Kremlin in Moscow. The Kremlin includes several buildings surrounded by a wall.

This gate is one entrance to the Kremlin in Moscow.

St. Petersburg is Russia's second largest city. It was Russia's capital for 200 years. The city has had two other names, Petrograd and Leningrad.

Russia is a country with more than 100 different ethnic groups. They have different languages and cultures. Most people in European Russia belong to the **Russian Orthodox Church**. There are also many Muslims in parts of Siberia.

The Soviet Union launched many rockets. Today Russia continues to send rockets into space.

Mikhail Gorbachev

Russia's History

From 1922 until 1991, Russia was part of a larger nation called the Soviet Union. The Soviet Union had 15 **republics**, or states. Russia was the largest and most powerful republic. A Communist government in Moscow ruled the entire nation. The Soviet Union was ruled by dictators.

The Communist leaders controlled the way people farmed. The government, not the people, owned most farms. Farmers were allowed to own very small **plots**, or areas of land. They could only grow crops for their own families on these plots. But one fourth of the nation's crops came from these small private plots.

The Soviet Union became an industrial country. Its Communist government owned all factories and businesses. The government did not allow Soviet factories to make many consumer goods. Instead the factories were used for **heavy industries**. Heavy industries make steel, machines, weapons, and other large products.

Communism helped the Soviet Union become a modern nation. The Communists sent rockets into space. They built a powerful army. People had more food. All children went to free schools. But people had no freedom in the Soviet Union. They could not move from one city to another without permission from the government. People were not allowed to practice their religion. People who spoke out against the Communist government were punished. Many millions were sent to jails in Siberia. Other people were killed. Many people were unhappy living with communism. They wanted more freedom.

In 1985 Mikhail Gorbachev became the Soviet Union's last Communist president. He allowed people to own businesses. Gorbachev also allowed much more freedom. This idea of openness and freedom was called by its Russian name, **glasnost**. With the new glasnost, people also wanted to be free from the control of the Communist leaders in Moscow. Some of the Soviet republics decided to become independent. At

Russia still has shortages of many goods. Here people wait in line to buy bread.

the end of 1991, the Soviet Union fell apart as a nation. Russia and the other republics became independent countries. Boris Yeltsin became Russia's president. Russia was no longer a Communist country.

Russia Today

Now that communism has ended in their country, Russians are enjoying more freedom. Now people can practice their religion. Privatization of stores, factories, and farms is taking place. But Russia has many difficult problems today. Most people earn low salaries. Many people do not have jobs. There is also a **shortage** of housing. There are not enough apartments in the cities for all the people who need them. There are also food shortages. People must stand in line for hours and pay high prices to buy food.

Inflation is another problem. The Communists had kept the prices of certain goods low so all people could buy them. Prices are now very high because the government does not control them. People find it hard to pay for expensive food and clothing.

Health care in Russia is very poor. The country does not have enough medicine to treat sick people.

There are **social problems**, too. There is much crime in Russia. **Alcohol abuse** is another problem. Many Russians drink too many alcoholic drinks.

Boris Yeltsin

The fighting in Chechnya has killed many people and destroyed many buildings.

A city in Chechnya during the fighting

There have also been problems with ethnic groups. In one part of Russia called Chechnya, the people want to have their own independent country. In 1994 they started fighting the Russian Army. They fought for almost two years. The fighting has stopped. No one knows what Chechnya's future will be.

Why are there so many problems in Russia? One reason is that the country does not have modern technology. It needs better technology for its farms, factories, and hospitals. Also Russia does not have enough transportation. Crops often rot in the fields because there is no way to get them to markets. Russia needs better ways to move resources and food.

Today Russia is trying to be a democracy. It is trying to have a free market economy. Russians need a higher standard of living. Then they will be able to enjoy life without communism.

Chapter Main Ideas

1. Plains cover most of Russia. Snow covers most of the country for at least six months of the year.
2. Communism ended in 1991. Since then, there have been food and housing shortages.
3. Siberia has most of Russia's resources. But Russia has not been able to use many of those resources.

Siberia

Siberia is the huge Asian part of Russia. This cold region has three fourths of Russia's land but less than one fifth of its people.

At one time Siberia had many jails. Communist leaders sent people who spoke against the government to those jails. Many people were in Siberian jails for more than twenty years.

There are many long rivers in Siberia that flow into the Arctic Ocean. The Russians have built dams on these rivers. These dams make electricity that is used in western Russia.

Lake Baikal is one of the most interesting places in Siberia. It is the world's deepest freshwater lake. It is more than 6,000 feet deep. Almost 2,000 different kinds of plants and animals live in Lake Baikal. At times the waves in the lake are 15 feet high.

Russia wants to develop Siberia since it has most of the country's natural resources. People who move to work in Siberia are paid salaries that are higher than salaries in other places in Russia.

The Trans-Siberian Railroad has made it easier for people and resources to travel through southern Siberia. Pipelines carry oil and gas to western Russia. But there are few roads and railroads in the north. New ways are needed to reach the rich resources of northern Siberia.

Lake Baikal in Siberia

Write a sentence to answer each question.

1. **Place**   Why is Lake Baikal an interesting place in Siberia?

2. **Movement**   How do people and resources move from Siberia to western Russia?

3. **Region**   What kind of region is Siberia?

4. **Human/Environment Interaction**   How have dams on Siberia's rivers helped Russia?

5. **Location**   What is Siberia's location?

◆ Vocabulary

Finish the Paragraph   Use the words in dark print to finish the paragraph below. Write the words you choose on your paper.

heavy industries   **tundra**   **alcohol abuse**
steppes   **plots**   **shortage**

Northern Siberia is covered with cold icy land called __1__. Southern Siberia has grassy plains called __2__. Russia's best farmland is there. In the past Russian farmers grew crops on small areas of land, or __3__. Russia has become an industrial country. Most Russian factories manufacture goods like steel, machines, and weapons. These are products of __4__. There are not enough apartments in Russian cities, so there is a __5__ of housing. One big social problem in Russia is __6__, which means people drink too many alcoholic drinks.

◆ Read and Remember

Where Am I?   Read each sentence. Then look at the words in dark print for the name of the place for each sentence. Write the name of the correct place on your paper.

Chechnya   **St. Petersburg**   **Black Sea**
Moscow   **northern Siberia**   **Trans-Siberian Railroad**

1. "I am in a port on the Baltic Sea."

2. "I am in Russia's capital."

3. "I am at a beach in southern Russia."

4. "I am in a region covered with tundra."

5. "I am in a region where fighting began in 1994."

6. "I am on a train traveling from the Pacific Ocean to Moscow."

◆ Think and Apply

Drawing Conclusions   Number your paper from 1 to 5. Read each pair of sentences. Then look in the box for the conclusion you might make. Write the letter of the conclusion on your paper.

1. Russia is on two continents.
Russia is almost twice the size of the United States.

Conclusion: _____

2. Snow covers much of Russia for six months each year.
Trucks travel on frozen rivers in the winter.

Conclusion: _____

3. Few people want to work or live in northern Siberia.
There are no railroads in northern Siberia.

Conclusion: _____

4. People who spoke out against the Soviet Union's leaders were sent to jail.
Communist leaders did not allow freedom of religion.

Conclusion: _____

5. There are shortages of housing and food in Russia today.
There are many people who do not have jobs.

Conclusion: _____

Conclusions
   A. There was little freedom in the Soviet Union.
   B. It has been difficult for Russia to change to a free market economy.
   C. Most of Russia has a very cold climate.
   D. Russia has not developed the resources of northern Siberia.
   E. Russia is a huge country.

◆ Journal Writing

Write a paragraph that tells three ways in which Russia has changed since communism ended.

Reading a Statistics Table

A **table** is a chart that has **statistics**, or numbers, that provide information about a topic. Tables can be used to compare, to contrast, or to draw conclusions. The table below gives statistics for the population and the number of cars in six countries. To read this table, first read the name of each heading. To find information about each heading, read the table from top to bottom. To find information about each country, read the table from left to right.

Read the statistics table. Then write on your paper the words in dark print that finish each sentence.

> **two**   **command economy**   **the United States**   **population**
> **eight**   **smallest**   **largest**

1. The United States has the _____ population of the six countries.

2. Ukraine has the _____ number of cars.

3. Russia has a larger population than every country on the table except for _____.

4. Poland and Spain have almost the same _____.

5. Spain has almost _____ times more cars than Poland.

6. France has about _____ times more cars than Ukraine.

7. We can conclude that countries that had a _____ have fewer cars.

The Population and Number of Cars for Six Countries

Country	Population	Number of Cars
Russia	148,200,000	11 million
United States	265,600,000	146 million
Poland	38,600,000	7 million
Spain	39,200,000	13 million
Ukraine	50,800,000	3 million
France	58,000,000	24 million

The Commonwealth of Independent States

Where Can You Find?
Where can you find a large port on the Black Sea?

Think About As You Read

1. Why was the C.I.S. formed?
2. How do C.I.S. countries differ from each other?
3. Why is Ukraine a poor country when it is rich in natural resources?

New Words

- organization
- nuclear weapons
- mosques
- radioactive wastes
- contaminated
- chernozem

People and Places

- Belarus
- Moldova
- Minsk
- Caucasus Mountains
- Caspian Sea
- Georgia
- Armenia
- Azerbaijan
- Central Asia
- Kiev
- Dnepr River
- Odessa
- Chernobyl

Imagine traveling through the 12 countries that are part of the Commonwealth of Independent States, the C.I.S. As you cross the border to go from one nation to another, you must show your passport to the border guards. You have to change your money to the currency of each republic that you visit. You also must know about the laws of each republic. Since the Soviet Union broke apart in 1991, each republic has its own laws, army, money, stamps, and flag.

Understanding the C.I.S.

The C.I.S. is an **organization**, or group, of 12 new countries. These countries had been republics of the Soviet Union. Russia is the largest C.I.S. country. Estonia, Latvia, and Lithuania are three republics that had been part of the Soviet Union, but did not join the C.I.S. These Baltic countries have closer ties with Western Europe than with Russia.

The C.I.S. was formed from 12 countries that had been part of the Soviet Union. What are the two largest countries in the C.I.S.?

THE COMMONWEALTH OF INDEPENDENT STATES

The C.I.S. was formed in 1991 to help the 12 new countries work together. One goal of the C.I.S. is to improve trade among these countries. Another goal is to control the **nuclear weapons** in this region. These weapons are powerful bombs that can destroy large cities. Russia and three other republics had nuclear weapons. These new nations gave Russia their nuclear weapons. Russia has promised to destroy the weapons.

Three Groups of Nations in the C.I.S.

The twelve C.I.S. countries can be divided into three regions. The first region includes Russia and three countries that are east of Russia—Ukraine, Belarus, and Moldova. Minsk, the capital of Belarus, is also the capital of the C.I.S. These four countries are the most industrial countries in the C.I.S.

The second region includes the three countries in the Caucasus Mountains. They are located between the Black Sea and the Caspian Sea. In two countries, Georgia and Armenia, most people are Orthodox Christians. In the third country, Azerbaijan, most people are Muslims. Georgia and Armenia have good farmland. Azerbaijan has oil, but it has not developed this resource yet.

This mosque is one of many found in Central Asia today.

The third region includes the five countries in Central Asia. In all of these countries, most people are Muslims. They feel closer to the Muslim countries of Asia than to the countries of Europe. There were not many **mosques** in Central Asia when it was part of the Soviet Union. A mosque is where Muslims pray. Since communism has ended, more people practice their religion. There are now thousands of mosques in Central Asia.

Central Asia has a dry climate. Many people work at farming or herding animals. These are the poorest countries of the C.I.S. Some of these countries have oil, but they have not developed that resource.

Each country in the C.I.S. has its own culture. Let's look more closely at Ukraine.

Ukraine: Its Geography, People, and History

Most of Ukraine is covered by the North European Plain. Ukraine has steppes with good farmland. Ukraine, like other countries of this region, has a continental climate. Winters are cold and summers are warm. Ukraine has a warmer climate than Russia. In the south, near the Black Sea, the climate is warmest.

Many people work at herding in Azerbaijan.

About 51 million people live in Ukraine. Two thirds live in cities. Kiev is the capital and the largest city. It is on the Dnepr River. The river is important for transportation and hydroelectric power. Odessa is a large, important port on the Black Sea.

Ukraine has a long history. But for hundreds of years, it was ruled by other countries. Ukraine was one of the first countries forced to become part of the

Ukraine has several large rivers. The Dnepr River flows into the Black Sea. What port city is located on the Black Sea?

Flag of Ukraine

The Chernobyl nuclear power plant

Soviet Union. The Soviet Union tried very hard to turn Ukraine into a Russian country. Only about one fifth of the people are Russians. All schools, television shows, and newspapers had to use the Russian language. But the people of Ukraine kept their own language and culture. In 1991 Ukraine became independent. It joined the C.I.S. that same year. Today Ukrainian is the official language. Many people speak both Ukrainian and Russian.

In 1986 a nuclear power plant in the city of Chernobyl caught on fire and exploded. The power plant was near Kiev. The plant had been built by the Soviet Union. The explosion sent large amounts of **radioactive wastes** into the air, water, and soil. Neighboring countries were polluted with these wastes. Ukrainians said at least 6,000 people died because of the accident. People continue to become sick from wastes in the air, the water, and the soil. The area around Chernobyl is still **contaminated**, or polluted, with these dangerous radioactive wastes.

Ukraine's Resources and Economy

Ukraine is rich in resources. Its fertile black soil is one of its most important resources. This soil, called

chernozem, gives Ukraine some of the best farmland in the world. About one fifth of the people are farmers. The country produces more wheat than any other country in Europe. Farmers also grow corn, rye, potatoes, and sugar beets.

Ukraine also has many minerals. It has large amounts of coal and iron for its industries. One fourth of all manufactured goods in the C.I.S. comes from Ukraine. Factories produce steel, machines, cement, glass, trains, and other products. But they do not make enough consumer goods. Ukraine imports consumer goods from Europe, North America, and Asia. Ukraine has some oil and natural gas. But Ukraine does not have enough for all its needs. It imports oil and natural gas from Russia and other C.I.S. countries. Ukraine trades other goods with Russia and the other C.I.S. nations.

Ukraine has some of the best farmland in the world.

Ukraine is moving toward a free market economy. More and more factories are now owned by private companies. But farms are still owned by the government. There are very few private farms.

Ukraine is rich in resources, but it is a poor country. Ukraine is a poor country because it lacks modern technology. Ukraine's factory products are not as good as those from Western Europe and the United States. Most workers earn only about $2,000 a year. Ukraine is working hard to improve its economy.

No one knows what the future of the C.I.S. will be. Will it become like the European Union and join the countries of the region together? Will the 12 members continue to be part of the C.I.S.? People around the world are watching to see how this region will change.

The people of Ukraine have a long history with many traditions.

Chapter Main Ideas

1. The C.I.S. was formed from countries that had been republics of the Soviet Union.
2. The 12 countries of the C.I.S. include many languages, ethnic groups, and cultures.
3. Ukraine produces large amounts of farm crops and factory goods. More modern technology will help improve Ukraine's economy.

◆ Vocabulary

Finish Up   Choose the word or words in dark print that best complete each sentence. Write the word or words on your paper.

> **nuclear weapon**   **radioactive**   **contaminated**
> **organization**   **mosque**   **chernozem**

1. An atomic bomb is a type of _____.

2. The fertile soil in Ukraine is called _____.

3. The waste products from making nuclear energy are _____ wastes.

4. The C.I.S. is a group, or _____, made from new countries that had been Soviet republics.

5. Water is _____, or not pure, when there are wastes in it.

6. A building where Muslims pray is a _____.

◆ Read and Remember

Write the Answer   Write one or more sentences to answer each question.

1. What are two goals of the C.I.S.?

2. Which three republics of the Soviet Union did not join the C.I.S.?

3. Which four C.I.S. countries are the most industrial?

4. Which group of C.I.S. countries has a large Muslim population?

5. What happened at Chernobyl?

6. What products does Ukraine manufacture?

7. Why is Ukraine a poor country?

8. What two religions are practiced in the three countries in the Caucasus Mountains?

9. Why are there many more mosques in the central Asian C.I.S. countries today than there were under communism?

10. Where is the port city of Odessa, Ukraine, located?

11. What did the Ukrainians do to show they were not Russians?

12. Why did the Chernobyl accident kill so many people?

13. Why is the Dnepr River important to Ukraine?

14. What important natural resources must Ukraine import?

◆ Think and Apply

Fact or Opinion   Number your paper from 1 to 10. Write **F** on your paper for each fact. Write **O** for each opinion. You should find four sentences that are opinions.

1. There are 12 republics in the C.I.S.

2. The C.I.S. will fall apart.

3. The Baltic countries should join the C.I.S.

4. Each republic has its own language, money, laws, and culture.

5. The republics of Central Asia have a dry climate.

6. Russia has more people than Ukraine.

7. Most of Ukraine's trade should be with the C.I.S.

8. Ukraine has fertile soil and mineral resources.

9. Ukraine farmers should own their own farms.

10. The explosion at Chernobyl caused serious pollution problems.

◆ Journal Writing

Imagine you are traveling through the C.I.S. Write a paragraph in your journal that tells which two countries you would visit. Tell why you would visit those countries.

Understanding a Resource Map

Resource maps show where resources can be found in an area. Some resource maps can show where minerals are found. Other maps show where farm and factory products are found. The resource map on this page shows where resources can be found in Ukraine.

Use the map key to find out which resources are shown. Then write the answer to each question on your paper.

1. What resource is near the Donets River?

coal   iron ore   oil

2. What two resources are found in southern Ukraine?

natural gas and coal   natural gas and iron ore   coal and oil

3. What resource is not found in western Ukraine?

coal   iron ore   oil

4. Where can you find natural gas?

Kharkiv   Donetsk   Odessa

5. What city would have many factories because it is near coal fields?

Odessa   Donetsk   Kherson

CHAPTER 22

The New Nations of Yugoslavia

Where Can You Find?
Where can you find a city where Olympic Games were played in 1984?

Think About As You Read

1. When did Yugoslavia first become a nation?
2. How did the end of communism change Yugoslavia?
3. Why did Bosnia have a civil war?

New Words

- Islam
- nationalism
- majority
- ethnic cleansing
- cease-fire
- central government

People and Places

- Sarajevo
- Yugoslavia
- Balkan Peninsula
- Croats
- Serbs
- Slovenia
- Croatia
- Serbia
- Montenegro
- Macedonia
- Bosnia and Herzegovina
- Bosnia
- Josip Broz Tito
- Dayton, Ohio

In 1984 many people came to the city of Sarajevo in Yugoslavia to watch and compete in the winter Olympic Games. At that time no one knew that within ten years Sarajevo would be destroyed by war.

The Location of Yugoslavia

The Balkan Peninsula is in southeastern Europe. Greece and several Eastern European countries are on this peninsula. Mountains cover most of the countries on the Balkan Peninsula. Most of this region has a continental climate. People near the coast enjoy sandy beaches and a Mediterranean climate.

Yugoslavia is one of the countries on the Balkan Peninsula. Since 1991 Yugoslavia has been changing. It has also been having a civil war.

Yugoslavia's History

When World War I ended, Europe's leaders created new borders between their countries. In some places they also created new countries. Several small countries were joined together to form Yugoslavia.

Four republics became new countries in this region when they left Yugoslavia. What two republics are still in Yugoslavia?

A Muslim in Bosnia

Several ethnic groups lived in these countries. Croats, Serbs, and Muslims were the three largest ethnic groups in Yugoslavia. These groups have lived in this region for hundreds of years. But they have often fought each other. Each group has a different culture and follows a different religion. Croats are Roman Catholics, and Serbs are Orthodox Christians. Muslims follow **Islam**, the world's second largest religion. Today there are more than one billion Muslims, mostly in Asia and Africa.

After World War II, Yugoslavia became a Communist country. At that time the country had six states, or republics. They were Slovenia, Croatia, Serbia, Montenegro, Macedonia, and Bosnia and Herzegovina. Bosnia and Herzegovina is often called Bosnia. Josip Broz Tito was Yugoslavia's Communist dictator. Tito forced the different ethnic groups to live together. After Tito died the ethnic groups began fighting again.

Communism ended in Yugoslavia in 1990. Then Croatia, Slovenia, Macedonia, and Bosnia became independent countries. In 1992 only the republics of Serbia and Montenegro remained part of Yugoslavia.

Much of Sarajevo has been destroyed because of the civil war in Bosnia.

War in Bosnia

Nationalism soon became a big problem in this region. Nationalism means a very strong love for your country or ethnic group. Nationalism was especially strong in Serbia. That country wanted all Serbs in the region to form a larger Serbian nation. They wanted to control the areas in Bosnia where Serbs lived.

A civil war began in Bosnia in 1992. It became Europe's worst war since World War II. Bosnia's ethnic groups fought to control the country. The Bosnian Serbs fought against the Muslims. The Serbs were angry that Muslims controlled Bosnia's government. Bosnian Croats also fought to control part of the country. Not one of these ethnic groups had a **majority**, or more than half, of Bosnia's people.

During the civil war, Bosnian Serbs captured about two thirds of Bosnia. Serbian soldiers from Yugoslavia helped the Bosnian Serbs. The Serbs removed and often killed all people from other ethnic groups who lived in areas the Serbs had captured. Thousands of people were killed. People have called this **ethnic cleansing**. Some Muslims and Croats also did some ethnic cleansing. Bosnia's ethnic cleansing reminded people of the killing of Jews in the Holocaust.

Leaders in France, Great Britain, Russia, and other European countries tried to end the war. They met

Ethnic Groups in Bosnia

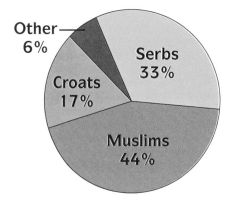

Other 6%
Croats 17%
Serbs 33%
Muslims 44%

NATO soldiers from Great Britain help carry out the peace plan in Bosnia.

with leaders from Bosnia, Croatia, and Serbia. These leaders agreed to a **cease-fire**, an agreement to stop fighting. But the fighting started again.

In 1995 the United States helped end the civil war. The leaders of Bosnia, Croatia, and Serbia met in Dayton, Ohio. They agreed to divide Bosnia into two states. Bosnian Muslims and Croats control one state. Bosnian Serbs control the other state. Each state has its own government. A **central government** in Sarajevo leads all of Bosnia. The government in Sarajevo is led by a Muslim, a Croat, and a Serb. These leaders take turns working as president of Bosnia.

Leaders in the United States and Western Europe feared that fighting could begin again in Bosnia. So they sent 60,000 NATO soldiers to Bosnia to carry out the peace plan. Thousands of American soldiers were sent to Bosnia in 1996. So far, NATO soldiers have kept peace in Bosnia.

Looking at Bosnia's Future

Bosnia's long civil war destroyed much of the country. The war destroyed much of Sarajevo and other cities. Many factories, farms, roads, and railroads are gone. Before 1990 many tourists visited Yugoslavia and its neighbors each year. The region no longer earns money from tourism.

Children in Bosnia have gone back to school now that the civil war has ended.

The Bosnians have begun to rebuild their country. They are slowly returning to work. Yugoslavia, Bosnia, and Croatia have many natural resources that could help their people have a higher standard of living.

The ethnic groups of this region continue to hate each other. Will people in Bosnia live in peace after NATO soldiers return home? Will another civil war begin? No one knows what Bosnia's future will be.

Chapter Main Ideas

1. Yugoslavia was created after World War I.
2. Communism ended in Yugoslavia in 1990. Four republics became independent countries. By 1992 only Serbia and Montenegro were part of Yugoslavia.
3. Bosnia's three ethnic groups fought a long civil war.

◆ Vocabulary

Match Up   Finish the sentences in Group A with words from Group B. Write on your paper the letter of each correct answer.

Group A

1. Muslims follow the religion of _____.

2. An agreement to stop fighting is a _____.

3. The main government that has control over other governments in a country is the _____.

4. A strong love for your nation or ethnic group is _____.

Group B

A. Islam

B. central government

C. nationalism

D. cease-fire

◆ Read and Remember

Complete the Geography Organizer   Copy the geography organizer shown below on your paper. Complete it with information about the new nations of Yugoslavia.

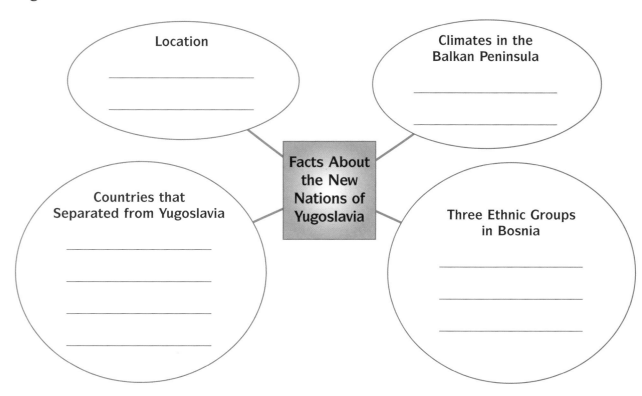

Matching   Each item in Group A tells about an item in Group B. On your paper write the letter of each item in Group B next to the correct number.

Group A

1. These people live in Croatia and are Roman Catholics.

2. Yugoslavia and Greece are located on this part of southeastern Europe.

3. He was the Communist dictator of Yugoslavia after World War II.

4. By 1992 only Montenegro and this republic remained part of Yugoslavia.

5. The 1984 Olympic Games were held in this city, but now the city has been destroyed by civil war.

Group B

A. Sarajevo

B. Balkan Peninsula

C. Croats

D. Serbia

E. Josip Broz Tito

◆ Think and Apply

Sequencing   Number your paper from 1 to 5. Write the sentences to show the correct order.

Yugoslavia became a Communist nation after World War II.

Yugoslavia became a nation after World War I.

NATO troops went to Bosnia to carry out the peace plan.

In 1992 a civil war began between Bosnia's ethnic groups.

Leaders of Bosnia, Croatia, and Serbia signed a peace plan in Dayton, Ohio.

◆ Journal Writing

Write a paragraph in your journal that tells why Bosnia's civil war began. Then tell how the United States helped to end the war.

Comparing Historical Maps

A **historical map** shows how a region looked during a certain period of history. You can learn how a region changes over time by comparing two maps of the same place from different time periods.

The historical map on the left shows Yugoslavia in 1945. It had six republics at that time. The map on the right shows Yugoslavia in 1992. In that year it had only two republics.

Study and compare the maps. Then use the words in dark print to finish the sentences. Write the sentences on your paper.

Romania	**Croatia**	**Serbia**
Hungary	**Albania**	**Adriatic Sea**

1. The two republics of Yugoslavia in 1992 were Montenegro and _____.

2. A country that was west of Yugoslavia in 1992 was _____.

3. A country that was south of Yugoslavia in 1945 and 1992 was _____.

4. A country that was north of Yugoslavia in 1945 and 1992 was _____.

5. A country that was east of Yugoslavia in 1945 and 1992 was _____.

6. Yugoslavia had a larger border on the _____ in 1945.

Yugoslavia in 1945

Yugoslavia in 1992

CHAPTER 23

Looking at Eastern Europe and Russia and Its Neighbors

Where Can You Find?
Where can you find a large sea that grows smaller each year?

Think About As You Read

1. What changes have taken place in this region since communism ended?
2. How can nations solve the problems of shortages and not enough technology?
3. How has pollution hurt this region?

New Words

- criticize
- political unrest
- birth defects
- irrigate

People and Places

- Bulgaria
- Hungary
- Aral Sea

Think of the many ways Eastern Europe and Russia and its neighbors have changed since communism ended. Newspapers now have articles that **criticize**, or complain about, the governments. Churches are filled with people on Sunday mornings. People are enjoying more freedom than ever before. But the countries in this region must solve six big problems so that people can have better lives.

Low Standards of Living, Many Shortages, and Not Enough Technology

There are six major problems in this region. The first problem is that most people in this huge region have a low standard of living. In many places people are poorer today than they were under communism. Governments no longer control the price of food and other goods. So products have become more expensive. People are forced to spend more of their money on the food they need.

Farmers in Bulgaria grow good crops, but they do not earn much money.

Some countries are much poorer than others. In Bulgaria many people earn only about $30 a month. In Poland workers may earn more than $300 a month. This is still much less money than an American worker would earn. Poland, Hungary, and the Czech Republic have the highest standards of living in the region. But people in these countries do not live as well as people in Western Europe.

The second problem is shortages. There are not enough apartments for all the people. Most countries do not have enough food. There are shortages of consumer goods because factories do not make enough of them. Communist governments had forced most factories to produce weapons and machines. Today these factories are producing more clothing and consumer goods. But there are still not enough goods for people who have money to buy them.

Shortages are a problem in this region. Shoppers in Budapest, Hungary, wait to buy goods.

Trade is helping the region get more consumer goods. Countries are importing consumer goods from Western Europe, Canada, the United States, and countries in Asia.

A third problem is the need for modern machines and new technology. Many people still use animals for farm work and transportation. Natural resources in Russia and in C.I.S. countries are wasted by old

Political unrest continues to be a problem in this region. Here 50,000 people speak out against their government in Serbia.

machines and old technology. To solve this problem, some countries are importing new technology.

Ethnic Problems, Pollution, and Political Unrest

Fighting between the different ethnic groups in a country is the fourth problem. In Chapter 22 you learned how fighting between ethnic groups in Bosnia led to a terrible civil war. In Russia and in the C.I.S., there have been many fights between ethnic groups. This problem can lead to more civil wars.

Political unrest is the fifth problem. Political unrest means people want to change their governments. After communism ended, countries held free elections. They elected people who were not Communists to lead their governments. But shortages and low standards of living have made people unhappy with their new governments. In some countries, such as Poland and Russia, many Communists have won elections to important government jobs again. Will Communist governments win control of many countries again? No one knows what will happen.

The last problem is the terrible pollution in this region. The Communist governments of Eastern Europe and the Soviet Union had tried to produce a lot of weapons, steel, and factory goods as fast as possible. So they did not care when factories sent pollution into the air. They allowed large amounts of waste to be dumped into rivers and seas. They chopped down forests and

Air pollution from industry is a big problem in this region.

did not plant new trees. So deforestation is a problem in many places.

Today air and water pollution are big problems in this region. A large number of children are born with problems, or **birth defects**, because of the pollution. The water of the Danube River is very polluted. In Russia's large city St. Petersburg, it is not healthy to drink the water. It is not even safe to take baths in St. Petersburg's water. Pollution problems are growing worse.

The Soviet Union hurt the environment near the Aral Sea in Central Asia. The Russians stopped river water from flowing into the Aral Sea. They used the river water to **irrigate** dry desert fields to produce crops. They grew cotton on those fields. Since the rivers did not flow into the Aral Sea, the sea became much smaller. Many plants and animals in the region have died. The desert around the sea has grown larger.

Ships are left in the desert sand as the Aral Sea gets smaller.

Looking at the Future

Eastern Europe and Russia and its neighbors is a region of rich resources. The nations are trying to use their resources to develop modern farms and factories. These countries have worked hard to change to free market economies. They are trying to allow more freedom. It will take time to solve the many problems. But if democracy continues in this region, people everywhere will enjoy a better life.

Chapter Main Ideas

1. Since communism ended, people are enjoying more freedom.
2. Countries changed to free market economies. This led to lower standards of living and many shortages.
3. Communists allowed factories to pollute the air and the water.

Pollution in this region hurts everyone.

◆ Vocabulary

Finish Up   Choose the word or words in dark print that best complete each sentence. Write the word or words on your paper.

political unrest   birth defects   criticize   irrigate

1. To bring water to dry land is to _____ the land.

2. Pollution has caused some babies to be born with problems, or _____.

3. When many people want to change the government, there is _____.

4. When you find fault with a government, you may _____ it.

◆ Read and Remember

Find the Answer   Find the sentences that tell about a problem in Eastern Europe, Russia, and the C.I.S. Write the sentences you find on your paper. You should find six sentences.

1. The region has a low standard of living.

2. There are shortages of food and consumer goods.

3. American companies have started businesses in some parts of Eastern Europe.

4. Eastern Europe is importing consumer goods from Western Europe.

5. There are many fights between ethnic groups.

6. Newspapers can criticize the governments.

7. Countries do not have modern machines and new technology.

8. There is political unrest in some countries in the region.

9. Communist governments have damaged the environment.

10. Some countries are importing new technology.

◆ Think and Apply

Find the Main Idea   On your paper copy the boxes shown below. Then read the five sentences. Choose the main idea and write it on your paper in the main idea box. Then find three sentences that support the main idea. Write them on your paper in the boxes of the main idea chart. There will be one sentence in the group that you will not use.

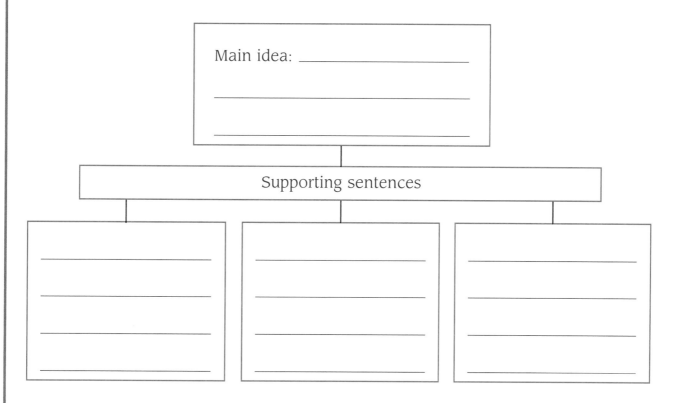

a. Factories dump wastes into the rivers and seas.

b. Eastern Europe and Russia and its neighbors have terrible pollution problems.

c. It is not safe to drink the water in many cities.

d. There is a shortage of apartments in Russia.

e. Factories send pollution into the air.

◆ Journal Writing

Eastern Europe and Russia and its neighbors have many problems. Choose two problems and write a paragraph about them in your journal. Then tell how people might solve the problems.

Reviewing Bar Graphs

You learned that **bar graphs** use bars of different lengths and colors to show facts. The bar graph below shows how much money was spent by four Eastern European countries to import goods in 1991 and 1994. Use the graph to compare their imports. Then circle the answer to finish each sentence.

1. The country with more imports in 1991 than in 1994 was _____.

    Poland   Russia   Hungary

2. The country with the most imports in 1991 and 1994 was _____.

    Poland   Hungary   Czech Republic

3. The Czech Republic spent _____ billion on imports in 1994.

    $26 $30 $13

4. Russia spent about _____ billion less on imports in 1994 than in 1991.

    $1 $8 $30

5. We can conclude that since communism ended, the amount of goods imported to these countries is _____ the amount of goods imported before 1991.

    greater than   lower than

    equal to

6. We can conclude that since Russia imported fewer goods in 1994, the number of shortages today is _____ under communism.

    larger than   smaller than

    the same as

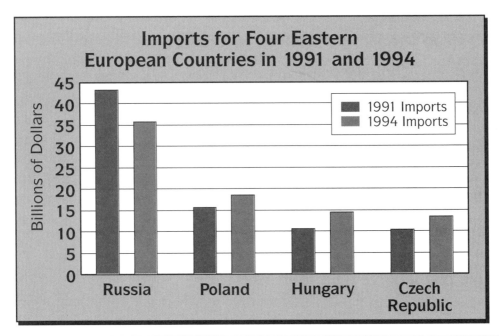

Imports for Four Eastern European Countries in 1991 and 1994

Africa

Victoria Falls

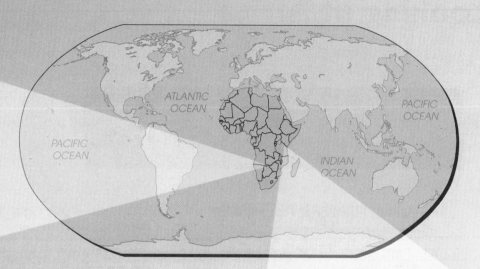

Mount Kilimanjaro

DID YOU KNOW?

- Africa has more countries than any other continent.

- The Sahara is a desert that is the size of the United States. It is the world's largest desert.

- Most of the world's gold and diamonds come from Africa.

- More than 800 languages are spoken in Africa.

- Victoria Falls drops about 350 feet. Africans call the falls *Mosi oa Toenja,* which means "smoke that thunders."

- The country of South Africa has three capital cities.

WRITE A TRAVELOGUE

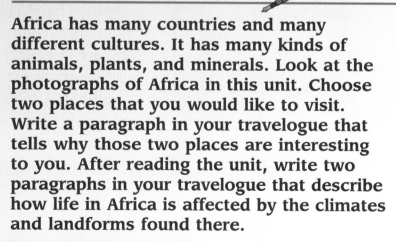

Africa has many countries and many different cultures. It has many kinds of animals, plants, and minerals. Look at the photographs of Africa in this unit. Choose two places that you would like to visit. Write a paragraph in your travelogue that tells why those two places are interesting to you. After reading the unit, write two paragraphs in your travelogue that describe how life in Africa is affected by the climates and landforms found there.

THEME: HUMAN/ENVIRONMENTAL INTERACTION

CHAPTER 24

Looking at Africa

Where Can You Find?
Where can you find a deep, wide crack in
Earth that is 4,000 miles long?

Think About As You Read

1. What are Africa's landforms and rivers?
2. What kinds of climates and plant life are found in Africa?
3. How is North Africa different from the region south of the Sahara?

New Words

- vegetation
- savannas
- semiarid
- Sahel
- grasslands
- droughts
- tribes
- traditional methods

People and Places

- Sahara
- Red Sea
- North Africa
- sub-Saharan Africa
- Atlas Mountains
- Mount Kilimanjaro
- Great Rift Valley
- Nile River
- Niger River
- Congo River
- Zambezi River
- Arabs
- Egypt
- South Africa

The continent of Africa has a lot of diamonds and gold. Africa is rich in many natural resources. But it is also one of the poorest regions in the world.

Africa's Landforms

Africa is located between the Atlantic Ocean and the Indian Ocean. Africa has a long, smooth coast. There are not many harbors or ports. So it has been hard for Africans to use the seas for trading.

The Sahara, which means "desert," stretches from the Atlantic Ocean to the Red Sea. It is the world's largest desert. Years can pass without rain in the Sahara.

The Sahara divides Africa into two regions. North Africa includes the Sahara and the region south of the Mediterranean Sea. The rest of Africa is called sub-Saharan Africa, or Africa below the Sahara.

A huge plateau covers most of Africa. The plateau is higher in eastern and southern Africa. A narrow coastal plain surrounds the plateau. The Atlas Mountains are

AFRICA

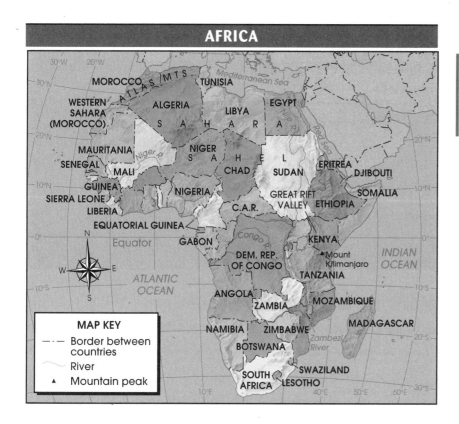

Africa has more nations than any other continent. What nation in Africa is also an island?

in northwest Africa. There are no mountain chains south of the Sahara. But there are some high mountains. The highest mountain in Africa is Mount Kilimanjaro. It is always covered with snow.

The Great Rift Valley is an important landform in east Africa. It runs north to south for more than 4,000 miles. The Great Rift Valley is made of deep, wide cracks in Earth's surface. Many lakes are in this valley.

Africa has four large rivers. The Nile River in east Africa is the longest river in the world. The Niger, Congo, and Zambezi rivers are other long African rivers. These rivers have many waterfalls. Waterfalls make it difficult to travel on the rivers. So it has been hard for Africans to travel and trade with each other. Waterpower from the rivers produces electricity.

The Great Rift Valley

Climate and Vegetation

Almost all of Africa lies in the tropics. But Africa has five climate regions. Each region has different **vegetation**, or plants such as trees, shrubs, and grass.

The region in central Africa near the Equator has a hot, wet tropical climate. There is heavy rain every day. Tropical rain forests grow in this climate. There

This part of the African savanna is called the Serengeti Plain.

is less rain as you move north or south away from the Equator. Areas at the Equator with high elevations have a cooler climate.

The second climate region is north and south of the rain forests. The climate is very hot, with both rainy and dry seasons. **Savannas** cover this region. Savannas are land areas with long, thick grass and short trees. Many wild animals live on the savannas.

North and south of the savannas, the climate is **semiarid**. It is still hot, but drier. There may be only ten inches of rain during the year. Short grasses grow in this third region. There are few trees. This semiarid region is part of the **Sahel**. The Sahel is a region of dry **grasslands** south of the Sahara. The Sahel stretches from the Atlantic Ocean to the Red Sea. Twelve countries are in the Sahel. The Sahel has long periods without rain called **droughts**. You will read more about this problem in Chapter 29.

The Sahel

The fourth region has a desert climate. The Sahara is north of the short grasslands. Smaller deserts are south of the southern grasslands at the southern end of Africa. The deserts have a few plants, but these plants need almost no water.

The fifth region has a Mediterranean climate. Summers are hot and dry. Winters are short and rainy. This climate is found in North Africa near the Mediterranean Sea. Grapes, olives, oranges, and other crops grow in this climate. This climate is also found at the southern tip of Africa.

Looking at North Africa

North Africa is the region that includes the Sahara and all the land north of it. North Africa has become very different from the rest of Africa. This is because the Sahara is so hard to cross. North Africa has the culture of the Middle East. Like the Middle East, most North Africans are Arabs. They speak the Arabic language. Most people are Muslims. Muslims follow the religion of Islam.

Egypt has the largest population in North Africa. It is one of the leaders of the Arab world. Most Egyptians live on the fertile land near the Nile River.

Oil is the most important resource in North Africa and the Middle East. You will learn more about North Africa and the Middle East in Unit 6.

Africa's History and Economy

People have lived in Africa for thousands of years. Early Africans did fine art work. They built large cities.

Hundreds of different ethnic groups have always lived in Africa. Usually these groups live together in close groups called **tribes**. Each ethnic group has its own language, religion, and culture.

From the 1800s to 1960, European countries ruled most of Africa. They wanted to own Africa's gold,

A mosque in Africa

The type of clothes that these women are wearing is part of their culture.

Sugarcane is an important crop in Egypt.

diamonds, copper, and other resources. Europeans divided most of Africa into colonies. Often Africans from different ethnic groups were forced to live together in one country. Many times these ethnic groups were enemies that could not get along. Today fighting between ethnic groups is still a big problem.

The Europeans needed modern transportation so they could mine Africa's resources. So they built roads, railroads, and seaports. They used these ports to ship minerals and raw materials to Europe. Raw materials are products from nature such as cotton, metal, and wood. They are used to make factory goods. Europeans often sold factory goods to their African colonies. Europeans also started plantations. They grew cash crops such as coffee, sugarcane, and cotton.

During the 1960s African nations began to rule themselves. They were no longer colonies. Now all African countries are independent.

Most Africans today earn a living by farming. They use **traditional methods** to grow crops. Most farmers use work animals to pull plows. They do not have modern farm machinery such as tractors. Farmers struggle to grow enough food for their families. Many people also raise sheep, goats, and cattle. Africans also work in mines. Africa exports its minerals to many nations. Each African nation earns most of its money by exporting only one or two farm or mineral products.

A gold mine in South Africa

Today all African countries except South Africa are developing nations. As you read this unit, notice how Africans are working to improve their countries.

Chapter Main Ideas

1. The Sahara divides Africa into two regions. North Africa has the people and the culture of the Middle East. Sub-Saharan Africa is a poor region that has rich resources.
2. Most of Africa has a hot climate, but there is more rain as you get closer to the Equator.
3. Most African nations became independent countries during the 1960s. All are now independent.

◆ Vocabulary

Analogies Choose the words in dark print that best complete the sentences. Write the word or words on your paper.

savannas   Sahel   drought   traditional methods   vegetation

1. Too much rain is to flood as too little rain is to _____.

2. Birds and fish are to animal life as trees and grasses are to _____.

3. Very tall trees are to tropical rain forests as long grasses are to _____.

4. Dry grasslands are to the _____ as rain forests are to tropics.

5. Heavy equipment is to modern technology as work animals are to _____.

◆ Read and Remember

Finish the Paragraph Number your paper from 1 to 8. Use the words in dark print to finish the paragraph below. Write the words you choose on your paper.

Equator   grasslands   waterfalls   Arab
Sahara   Sahel   Islam   plateau

The continent of Africa has more countries than any other continent. The __1__ divides Africa into two regions. It is the largest desert in the world. North Africa has the __2__ culture of the Middle East. North Africa borders the Mediterranean Sea. Most North Africans are Muslims who follow the religion of __3__. The area south of the Sahara is the __4__. It has dry __5__. A large __6__ covers most of the African continent. There are four large rivers in Africa. African rivers have __7__ that make transportation by boat dangerous. Most of Africa is hot. There is less and less rain as you move away from the __8__.

◆ Think and Apply

Cause and Effect Number your paper from 1 to 7. Write sentences on your paper by matching each cause on the left with an effect on the right.

Cause

1. Africa's long, smooth coast does not have many harbors, so _____.

2. The hot land near the Equator gets lots of rain, so _____.

3. It is difficult to travel across the Sahara, so _____.

4. Desert covers most of Egypt, so _____.

5. European nations needed raw materials for their factories, so _____.

6. Europeans needed transportation in order to mine Africa's resources, so _____.

7. During the 1960s, most African nations became independent, so _____.

Effect

A. they started colonies in Africa

B. people live on fertile land near the Nile River

C. tropical rain forests grow in the region

D. they are no longer colonies

E. Africa does not have a lot of trade with other parts of the world

F. there are great differences between northern and southern Africa

G. they built roads and railroads

◆ Journal Writing

You have been reading about the continent of Africa. You learned that the Sahara divides Africa into two regions. Imagine you are beginning a journey through Africa. Would you want to visit North Africa or Africa south of the Sahara? Write a paragraph in your journal telling which part of Africa you want to visit. Write two or more reasons to explain why you made this choice.

A **vegetation map** helps you learn which regions have different types of vegetation, or plant growth. Climate affects the type of vegetation that grows in a region. In Africa, vegetation changes where there is more rain or less rain. The map key tells you the color used on the map to show each type of vegetation.

Study the vegetation map and its map key on this page. Then finish each sentence in Group A with an answer from Group B. Write the letter of the correct answer on your paper.

Group A

1. You can find Mediterranean vegetation in North Africa near the Mediterranean Sea and at the _____.

2. You will not find a tropical rain forest in _____.

3. The largest dry vegetation region in North Africa has _____.

4. You will find grasslands in a hot, rainy climate to the north and south of the _____.

5. East of the tropical rain forest at the Equator are grasslands in a _____.

6. Africa's smallest vegetation regions have _____ vegetation.

Group B

A. North Africa

B. tropical rain forest

C. hot, wet climate

D. Mediterranean

E. tip of South Africa

F. desert plants

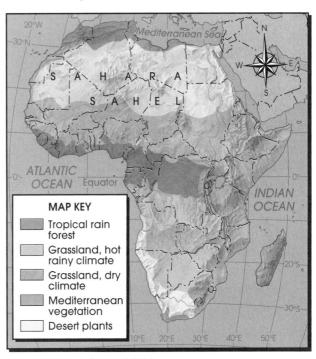

Vegetation Map of Africa

CHAPTER 25

Nigeria: The Oil-Rich Nation of West Africa

Where Can You Find?
Where can you find a lake that was formed by building a dam on the Niger River?

Think About As You Read
1. How is southern Nigeria different from northern Nigeria?
2. How does Nigeria earn its money?
3. What problems are found in Lagos?

New Words
- swamps
- delta
- reservoir
- yams
- cassava
- assemble
- urbanization
- overcrowded

People and Places
- Nigeria
- Jos Plateau
- Gulf of Guinea
- Benue River
- Kainji Lake
- Lagos
- Abuja
- Kano
- Hausas
- Yorubas
- Ibos

Nigeria is the only oil-rich country south of the Sahara. Nigeria also has more people than any other African country. About 104 million people live in Nigeria. Oil and a large population make Nigeria an important country. But it is still a developing country with millions of poor people.

Nigeria's Climates and Landforms

Nigeria is in western Africa. Much of Nigeria is on the Jos Plateau. Nigeria's southern coast is on the Gulf of Guinea. This gulf is part of the Atlantic Ocean.

Southern Nigeria has a hot, rainy tropical climate near the Equator. Coastal plains cover the land near the gulf. Farmers in the south grow cocoa beans, rubber trees, and palm trees. The area also has tropical rain forests and **swamps**. A swamp is soft, wet land.

Central Nigeria is in Africa's second climate region. Its climate is hot, with both wet and dry seasons. Savannas cover parts of central Nigeria.

NIGERIA

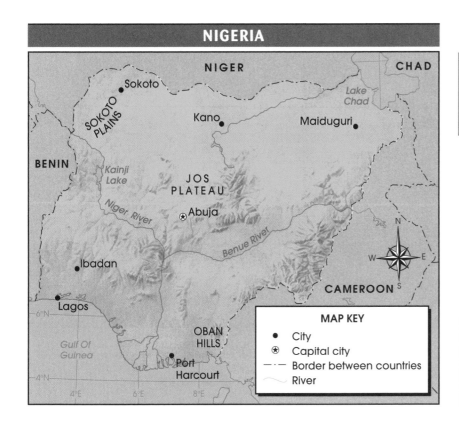

The Benue and Niger rivers flow through Nigeria. Where is the delta of the Niger river located?

Nigeria's flag

Northern Nigeria is part of the Sahel. Some places get almost no rain. Dry grasslands cover the north. Farmers in the north grow cotton, sugarcane, peanuts, and wheat. They also raise cattle, sheep, and goats.

There are many rivers and streams in Nigeria. In central Nigeria the Benue River flows into the Niger River. The Niger is the third largest river in Africa. It flows from western Nigeria toward the center of the country. Then it flows south into the Gulf of Guinea. The Niger River forms a large **delta** near the Gulf of Guinea. A delta is land made of soil and sand that the river has carried and left at the place where the river flows into the sea. Nigeria's oil is found near the delta.

In the 1960s, Nigerians built a large dam on the Niger River. The dam stopped the river from causing dangerous floods. The dam used river water to produce lots of electricity. Water from the river was also used to make a new lake called Kainji Lake. Kainji Lake is a **reservoir**, a place that stores water for future use. People use Kainji Lake for fishing. They also use water from the lake to irrigate dry land so they can grow crops.

Traveling on a swamp in Nigeria

Nigeria exports oil all around the world.

Nigerians study better ways to grow cassavas.

Nigeria's Resources and Economy

Nigeria has coal, gold, iron, and tin. But oil is the country's most important resource. Oil was discovered in Nigeria in the 1960s. Nigeria sells oil to the United States and to many other countries.

Nigeria depends on selling oil to earn money. The country has problems when the price of oil goes down. That happened in the 1980s. Nigeria had to sell its oil for a lower price. The country earned much less money. Since then, the price of oil has gone up. About 90 percent of the country's money now comes from exporting oil. This money from oil has helped only a few Nigerians. Most people are very poor.

Most Nigerians are subsistence farmers. They eat rice, beans, corn, and **yams**. They also eat an African vegetable called **cassava**. Most Nigerians eat very little meat. Most farmers live in small villages and use traditional methods of farming. So they do not grow much food. The government is trying to use oil money to help farmers grow more food. It has bought better seeds and fertilizers. But the country still does not grow enough food for its people. So Nigeria imports a lot of food. But Nigeria does export some cash crops such as cocoa, cotton, rubber, peanuts, and palm oil.

Only one fourth of Nigeria's people live in cities. Lagos is the largest city. It has more than one million people. It is a port on the Gulf of Guinea. Lagos has many factories. Some of the factory products are cement, clothing, steel, and food products. Some factories also **assemble** cars and radios. Factory workers use parts that are made in other countries.

Lagos is a crowded city with big problems. Each year more poor farmers leave their villages and move to Lagos. They hope to find factory jobs there. This movement of people to cities is called **urbanization**. Poor people in Lagos live in slums. The city does not have enough clean water, schools, houses, jobs, or transportation. There is a lot of crime in Lagos.

At one time Lagos was the capital of Nigeria. In 1980 the government began to build a new capital in central Nigeria. The new capital is Abuja. It is not yet finished, so many government offices are still in Lagos. Kano is the largest city in the north. Most houses in Kano are made of baked mud. Long ago, Kano was a center of trade for people who had crossed the Sahara.

History, People, and Government

People have been living in Nigeria for more than 1,000 years. Most Nigerians belong to three large

Kano, Nigeria

Lagos used to be the capital of Nigeria. Lagos is a very crowded city.

For hundreds of years, the people of Nigeria have used mud bricks to build homes and other buildings.

School children in Nigeria

ethnic groups. In the north many people are Hausas. They follow the religion of Islam. In the south many people are Yorubas and Ibos. Most people in these groups are Christians.

There have been problems among the country's many ethnic groups. There are more than 250 different groups. Each group has its own culture and its own language. It is hard for Nigerians to feel that they are part of one country. It is hard to feel like one country because so many different languages are spoken. People from different groups also find it hard to work together to solve problems.

During the late 1800s, Great Britain began to rule Nigeria. In 1960, Nigeria became an independent country. The new government made English the official language. It is the language used in Nigerian schools.

Nigeria is not a democracy. An army dictator rules Nigeria. A small group of rich people control the country. People are arrested if they work against the government.

Nigeria has many problems today. The population is growing very fast. Cities are **overcrowded** because they have too many people. There are not enough doctors and hospitals. There are not enough schools, so only half of the people can read and write. Farmers do not grow enough food for all the people. Most people have a very low standard of living. The country earns lots of money from its oil, but that money has not helped most people. Nigeria's leaders must find new ways to use the money from oil to help the Nigerian people.

Chapter Main Ideas

1. Nigeria has a hot, wet climate in the south. Northern Nigeria is hot and dry.
2. Nigeria earns most of its money from exporting oil. But most Nigerians have a low standard of living.
3. Nigeria has more than 250 ethnic groups and languages. English is the official language.

◆ Vocabulary

Finish Up   Choose the word in dark print that best completes each sentence. Write the word on your paper.

urbanization	**cassava**	**reservoir**
swamp	**delta**	**overcrowded**

1. A _____ is an area of soft, wet land.

2. A _____ is land made of soil and sand where a river flows into the sea.

3. A place where water is stored for future use is a _____.

4. A _____ is an African vegetable.

5. When many people move from villages to cities, there is _____.

6. Cities that have too many people are _____.

◆ Read and Remember

Where Am I?   Read each sentence. Then look at the words in dark print for the name of the place for each sentence. Write the name of the correct place on your paper.

Lagos	**Niger River**	**Gulf of Guinea**
Kano	**Abuja**	**Kainji Lake**

1. "I am on the third largest river in Africa."

2. "I am fishing at a lake that was formed by a dam on the Niger River."

3. "I am in Nigeria's southern port and largest city."

4. "I am in Nigeria's new capital."

5. "I am in a large city in northern Nigeria."

6. "I am on the body of water south of Nigeria."

◆ Think and Apply

Find the Main Idea   On your paper copy the boxes shown below. Then read the five sentences. Choose the main idea and write it on your paper in the main idea box. Then find three sentences that support the main idea. Write them on your paper in the boxes of the main idea chart. There will be one sentence in the group that you will not use.

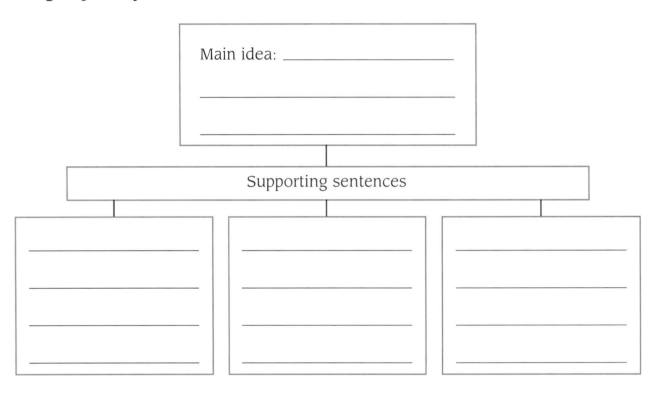

Main idea: _____

_____

_____

Supporting sentences

A. About half of Nigeria's people cannot read or write.

B. Most farmers use traditional methods.

C. English is Nigeria's official language.

D. Nigeria is a developing country with millions of poor people.

E. Only one fourth of the people live in cities.

◆ Journal Writing

Lagos has many problems because it is overcrowded. Write a paragraph that describes what Nigeria's government could do as it builds Abuja to avoid the problems it has in Lagos.

Reading a Climate Map

A **climate map** helps you learn about the weather in an area. It shows which places are rainy and which places are dry. A climate map helps you learn which places are hot and which are cooler. The map key on a climate map uses different colors to show different climates.

The map below shows the three climates found in Nigeria. As you travel from north to south in Nigeria, the climate becomes wetter.

Study the climate map. Then write the answer to each question.

1. What kind of climate does Lagos have?

2. What kind of climate is at the Niger River delta?

3. What climate does Abuja have ?

4. Is Kano's climate wetter or drier than that of Port Harcourt?

5. What two cities get less rain than Abuja?

6. What climate does Sokoto have?

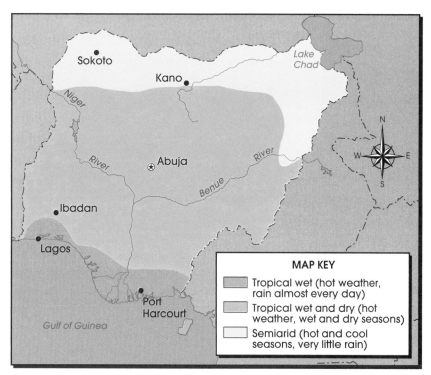

Climate Map of Nigeria

CHAPTER 26

Kenya: A Country in East Africa

Think About As You Read

1. What are Kenya's landforms and climates?
2. What products does Kenya export?
3. What is Kenya's biggest problem?

New Words

◆ nomads
◆ Swahili
◆ sisal
◆ wildlife
◆ national parks
◆ oil refineries

People and Places

◆ Kenya
◆ Mombasa
◆ Masai
◆ Mount Kenya
◆ Lake Victoria
◆ Kisumu
◆ Nairobi

Kenya is an African country that does not have silver or diamond mines. It does not have oil, coal, or iron. This nation has few resources. But Kenya has fertile land and beautiful places to visit.

Kenya's Landform Regions

Kenya is in eastern Africa. Find Kenya on the map on page 209. Notice that the Indian Ocean is to the east. The Equator passes through the middle of Kenya. What countries share borders with Kenya?

There are three landform regions in Kenya. There are plains near the Indian Ocean. These plains have a hot, wet climate. The soil is fertile and this region has some farming. Tropical rain forests and swamps are also in this plains region. There are beautiful beaches at the ocean. Mombasa is a busy port city on the Indian Ocean.

The second landform region is the plateaus that cover three fourths of Kenya. In most places these

KENYA

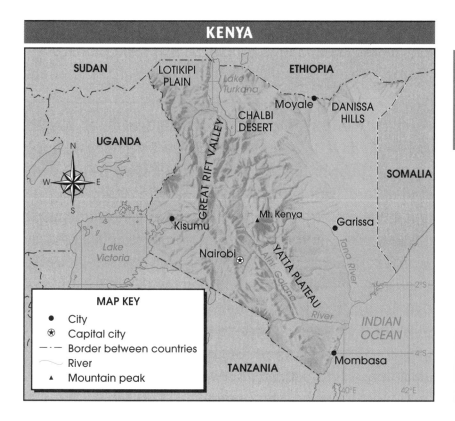

Kenya is located in eastern Africa. Mount Kenya has the highest elevation in Kenya. What lake is located on Kenya's western border?

Kenya's flag

plateaus get little rain. Some plateaus are covered with savannas. A small group of Kenyans called the Masai are **nomads**. The Masai move across the plateaus herding cattle. Thousands of wild animals also live on Kenya's plateaus.

The highlands are the third landform region. Hills and mountains cover this land in southwest Kenya. Mount Kenya is in the highlands region. Mount Kenya is more than 17,000 feet high. It is located at the Equator. But snow and ice always cover the top of the mountain because of its high elevation.

The Great Rift Valley cuts through the highlands region. Many lakes are part of this valley. Part of Lake Victoria, Africa's largest lake, is on Kenya's border. People do a lot of fishing in this area. More than 200 kinds of fish live in this huge lake. Kisumu, Kenya's third largest city, is on Lake Victoria.

The high elevation of the highlands gives this region a mild climate. The land around Lake Victoria gets a lot of rain. There is enough rain for farming. The Great Rift Valley has very fertile soil. There are many farms in this region. So most of Kenya's people live in the

Masai herding cattle

Nairobi, Kenya

rural highlands. Nairobi, Kenya's capital and largest city, is in the highlands. About 2 million people live in Nairobi. This modern city is about one mile above sea level.

Kenya's History, People, and Culture

People have lived in Kenya for thousands of years. In 1895 Great Britain took control of Kenya. The British started large farms for growing tea and coffee. They also built a railroad across the country. That railroad made it possible to travel from Mombasa to Nairobi. From Nairobi you could travel by train to Lake Victoria. The railroad also connected Kenya with other African countries.

In the 1950s the people of Kenya began to fight for their freedom from Great Britain. In 1963 Kenya became an independent country. Kenyans wanted an African language as their official language. **Swahili** became the country's official language. English is the country's second official language.

Almost all Kenyans are black Africans. But Kenya has some Asians and whites. About three fourths of all Kenyans are Christians. Some people are Muslims. Many people follow traditional African religions.

About 28 million people live in Kenya. Most of them live in small villages. They belong to about 40 different ethnic groups. Each group has its own language and customs. Kenya's government has been trying to help the many groups get along. But in 1993 there was terrible fighting between a few groups. Thousands of people were killed.

Earning a Living and Protecting the Wildlife

A sisal plantation

Most Kenyans earn a living in one of three ways. Most people work at farming on small farms. They grow subsistence crops such as corn, cassava, rice, and beans. They do not use modern ways of farming. But there are also large tea and coffee plantations where modern methods are used. Kenya earns more money from exporting coffee than from any other product. Kenyans also grow and export cotton and **sisal**. Sisal is a plant that is used to make rope.

Tourists can see zebras and other animals in Kenya's national parks.

Tourism is the second way Kenyans earn money. Tourists from many parts of the world visit Kenya each year. Thousands of Kenyans work in tourist hotels and restaurants. Many tourists enjoy Kenya's beaches and cities. Some spend days climbing Mount Kenya.

Other tourists come to Kenya to see **wildlife**, or many kinds of wild animals. Kenya has large **national parks**, parks owned by the government, where wild animals are protected. These parks cover hundreds of miles of land. Hunting is not allowed in the parks. Tourists drive through these parks to watch and to photograph the wildlife. Elephants, zebras, lions, and giraffes can be seen in Kenya's national parks.

The third way Kenyans earn money is by making factory goods. Kenya is a developing country. Kenya still buys most of the factory goods it needs from Western Europe. But it is building many new factories. Today Kenya's factories make chemicals, clothing, food products, cement, and paper. Some factories assemble cars. Kenya also has **oil refineries**. In an oil refinery, oil is cleaned and made into different products such as motor oil and gasoline. Kenya does not have oil, so its refineries use oil from other countries.

Mount Kenya

Looking at Kenya's Future

Kenya has several problems. About half of the people are poor. They have a low standard of living.

Mombasa is Kenya's busiest port. It is located on the Indian Ocean.

Most people do not own a television, a telephone, or a car. Kenya's biggest problem is rapid population growth. This means the population is growing very fast. Most families have six or more children. Kenya has one of the fastest growing populations in the world. But only one fifth of the land can be farmed. It will be hard for Kenyans to grow enough food for many more people. The country will also need more jobs. These are important problems for Kenya.

Kenya is working on these problems. It now has more farms, factories, and tourists than ever before. Most people can read and write. Each year new schools are started. Now Kenya must try to solve the problem of rapid population growth. Then people will enjoy a higher standard of living.

Chapter Main Ideas

1. Most of Kenya's people live in the southwest highlands. The soil in this region is good for farming.
2. Coffee is Kenya's most important export. Kenyans work at farming, tourism, and factory jobs.
3. Kenya's biggest problem is rapid population growth. Kenya does not have enough fertile land to grow food for a larger population.

Tourists in Kenya

◆ Vocabulary

Finish the Paragraph   Use the words in dark print to finish the paragraph below. Write the words you choose on your paper.

sisal   Swahili   national parks   oil refineries   nomads

The Masai are ___1___ who move from place to place. They herd cattle on Kenya's plateaus. Most people in Kenya live in the highlands. There they grow a plant for making rope called ___2___. They also grow tea and coffee. Kenyans speak an African language called ___3___. Tourists enjoy seeing Kenya's wild animals in large ___4___ owned by the government. Kenya has ___5___ where oil is cleaned so it can be made into products.

◆ Read and Remember

Complete the Geography Organizer   On your paper copy the geography organizer and complete it with information about Kenya.

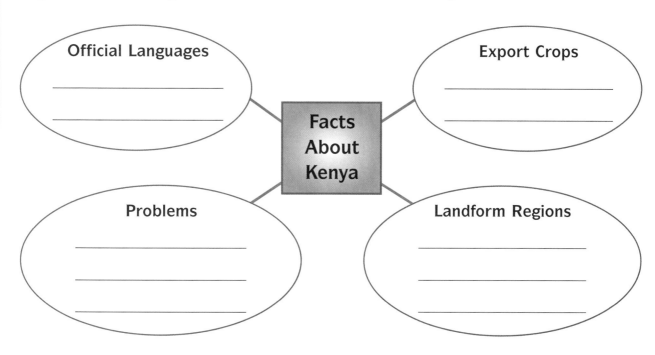

Official Languages
_____
_____

Export Crops
_____
_____

Facts About Kenya

Problems
_____
_____
_____

Landform Regions
_____
_____
_____

Find the Answer   Find the sentences that tell about a problem in Kenya. Write the sentences you find on your paper. You should find three sentences.

1. Many tourists visit Kenya each year.

2. Kenya has few natural resources.

3. The highlands have fertile soil.

4. Kenya's population is growing very fast.

5. Many people in Kenya are poor.

◆ Think and Apply

Compare and Contrast   Copy the Venn diagram shown below on your paper. Then read each phrase below. Decide if it tells about Nigeria, Kenya, or both. Write the number of the phrase in the correct part of the Venn diagram on your paper.

1. East African country
2. oil is the main export
3. many ethnic groups
4. 104 million people
5. Great Rift Valley

6. coffee is the main export
7. West African country
8. became independent from Great Britain
9. developing country

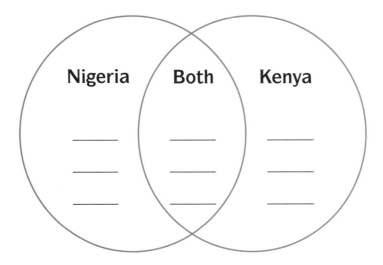

◆ Journal Writing

Write a paragraph in your journal that tells why rapid population growth is a big problem in Kenya.

Reviewing Line Graphs

The **line graph** on this page shows how Kenya's population has changed since 1970. A quick look at the graph tells you that the population has grown larger each year. Do you think Kenya's population next year will be larger than it is now?

Look at the graph on this page. Number your paper from 1 to 5. Then finish each sentence in Group A with an answer from Group B. Write the letter of the correct answer on your paper.

Group A

1. In 1970 Kenya's population was about _____.

2. In 1995 Kenya's population was about _____.

3. The graph does not show the year 1965. We can guess that Kenya's population in 1965 was _____ than in 1970.

4. The smallest population change was between 1970 and _____.

5. We can guess that in the year 2000, Kenya's population will be _____ than in 1995.

Group B

A. greater

B. smaller

C. 1975

D. 11 million

E. 28 million

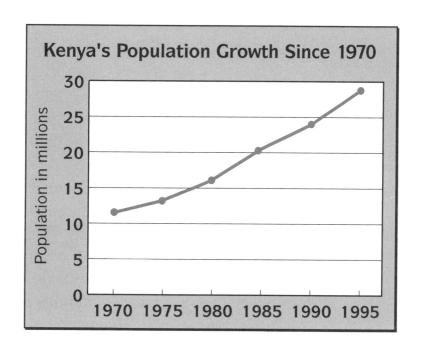

CHAPTER 27

Congo: The Largest Country in Central Africa

Where Can You Find?
Where can you find the world's longest lake?

Think About As You Read

1. How does the Congo River help Congo?
2. What problems are found in Congo's cities?
3. What resources are found in Congo?

New Words

- okapi
- Lingala
- malnutrition
- billionaire
- conflicts
- refugees
- rebelled

People and Places

- Inga Dam
- Congo Basin
- Lake Tanganyika
- Kinshasa
- Belgian Congo
- (Joseph) Mobutu Sese Seko
- Laurent Kabila
- Rwanda
- Hutus
- Tutsis

Congo should be one of the richest countries in Africa. It has lots of copper, diamonds, oil, and gold. Congo has rich resources. But it is one of Africa's poorest countries.

Congo's Landforms and Rain Forest

Congo is located in the central part of Africa. The Equator passes through the middle of Congo. All of Congo is in the tropics. Most of Congo is surrounded by other countries. Congo only has a 25-mile strip of land along the Atlantic Ocean.

The Congo River flows through most of the country. The river empties into the Atlantic Ocean. This river is almost 3,000 miles long. Many shorter rivers flow into the Congo River. People use the Congo River for fishing. People also travel on the river for hundreds of miles. In some places waterfalls make it impossible to travel on the river. The people of Congo built the Inga Dam on the Congo River. This dam makes large amounts of electricity.

DEMOCRATIC REPUBLIC OF CONGO

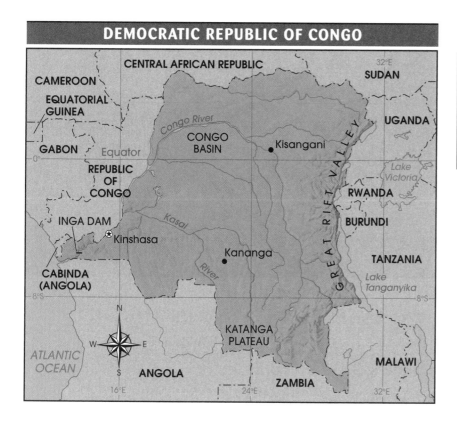

Congo is the largest country in central Africa. The Equator passes through this country. What city in Congo is located near the Equator?

The Congo Basin surrounds the Congo River. A basin is an area of land that is lower than the land around it. The Congo Basin is on a large plateau. A huge tropical rain forest covers much of the Congo Basin. This rain forest is the largest in Africa. Its trees are so thick and tall that the sun never reaches the floor of the forest. Thousands of different plants and animals live in the rain forest. The **okapi** is a furry animal that lives only in Congo's rain forests. It does not live in any other part of the world. The okapi is the symbol of Congo.

Congo's rain forest is being destroyed. People chop down trees to clear the land for farming. Trees from the forest are made into furniture and paper. People around the world want Congo to protect its rain forest from deforestation.

A plateau covers most of the land south of the Equator. Savannas cover this plateau. The climate is warm and rainy, but there is a dry season. The east and southeast have highlands. Hills and mountains cover this region. These highlands are part of the Great Rift Valley. Lake Tanganyika is in the Great Rift

An okapi

Kinshasa is a port on the Congo River.

Valley on Congo's eastern border. It is the world's longest lake. It is more than 400 miles long.

Congo's People and Cities

About 47 million people live in Congo. Most of the people of Congo are black Africans. About 80 percent are Christians. Congo has more than 200 ethnic groups and about 200 different languages. French is Congo's official language. **Lingala** and Swahili are African languages that many people use.

About 40 percent of the people live in cities. Kinshasa is the capital and largest city. It has almost 5 million people. It is on the Congo River.

Congo's cities have many problems. There are few jobs, so unemployment is a big problem. Cities have no buses or trains. There are not enough schools or hospitals. Poor people live in terrible slums. Congo's cities are also dangerous because there is lots of crime.

Many Resources But a Low Standard of Living

Congo has many minerals. It earns most of its money by exporting minerals. Its most important exports are copper and diamonds. Congo also has oil, gold, tin, silver, and cobalt. The country also exports some cash crops. Coffee, cocoa, tea, and cotton are sold to other countries.

Most people in Congo live in small rural villages. Their standard of living is very low. Most villages do not have even one car, television, or telephone. Few

Some people in Congo earn a living by fishing.

homes have running water or electricity. Most people are struggling subsistence farmers who use hand tools. Their main foods are corn, rice, and cassava. Many people suffer from **malnutrition** because they do not get enough to eat. Malnutrition is poor health that is caused by a lack of healthy food.

Although most people in Congo are poor, many children in Congo go to school. Most schools have few books and supplies.

Congo's History

In 1885 the central part of Africa became a colony of Belgium. Belgium is a small country in Europe. This colony was called the Belgian Congo. It included land that would later become Congo. Belgium earned a lot of money from mining the colony's metals and minerals. The Belgians built cities, schools, railroads, churches, and hospitals. They also brought their French language to the colony.

The Africans living in the Belgian Congo wanted to rule themselves. In 1960 the colony became a free nation. Then it was called the Congo. Today the country is called the Democratic Republic of the Congo. Most of the time it is just called Congo. Between 1960 and 1965, different people in Congo fought to win control of their new country. Finally in 1965 an army general named Joseph Mobutu became the president of most of Congo. He ruled as a dictator until 1997. During that time, Congo became a very poor country. It became poor because Mobutu stole his country's money and minerals. Mobutu became a **billionaire**.

Conflicts, or fights, between ethnic groups have always been a problem in Congo. In 1994 that problem grew worse when two ethnic groups began fighting in Rwanda. Rwanda is Congo's eastern neighbor. The two groups, the Hutus and the Tutsis, fought a war. During the war the Hutus murdered at least 500,000 Tutsi people. The war ended when the Tutsis won control of the government. Many Hutus left Rwanda. They became **refugees** in eastern Congo. Refugees are people who leave their country during a war.

A village in Congo

A rebel soldier in Congo during the civil war

Many people left Rwanda and became refugees in Congo.

Laurent Kabila

These refugees encouraged Congo's Hutus and Tutsis to fight against each other.

In 1996 fighting among ethnic groups became a civil war in Congo. Millions of people were unhappy with Mobutu. Laurent Kabila became the leader of a civil war against him. The Tutsis were the largest group that **rebelled**, or fought against, Mobutu. The Hutus fought for Mo@utu during the civil war. In 1997, after seven months of fighting, Kabila and the rebels won.

After the civil war, Kabila became president of Congo. He promised to improve the country's standard of living and help it become a democracy. But in 1997 Kabila said it was against the law to protest against the government. People could not join political parties. Many people wonder if Congo will improve without Mobutu. Will Congo use its resources to raise its standard of living? Will the country become a democracy? No one knows what the future of this mineral-rich country will be.

Chapter Main Ideas

1. Savannas cover the southern part of Congo. Northern Congo has a large tropical rain forest that is being destroyed.
2. From 1965 to 1997 President Mobutu ruled Congo as a dictator.
3. Laurent Kabila led Tutsi rebels against Mobutu and the Hutus during a civil war. The rebels won and Kabila became the new president.

◆ Vocabulary

Find the Meaning Choose the word or words that best complete each sentence. Write the words on your paper.

1. A **basin** is land that is surrounded by _____ land.

lower   drier   higher

2. An **okapi** is a _____ that lives only in Congo.

furry animal   fish   large cat

3. **Lingala** is an African _____.

town   city   language

4. **Malnutrition** means people have poor health because they do not get enough _____.

clothing   food   money

5. A **conflict** is a _____.

fight   mountain   culture

6. A **refugee** is a person who moves to another country because of a _____.

job   party   war

7. When people **rebelled**, they _____ against their government.

spoke   fought   wrote

◆ Read and Remember

Write the Answer Write one or more sentences to answer each question.

1. How do people use the Congo River?

2. What is happening to Congo's tropical rain forest?

3. What problems are found in Congo's cities?

4. What problems do people in Congo have because of the low standard of living?

5. Which ethnic group won the civil war in Rwanda?

6. Why did the Tutsis begin fighting against Congo's army in 1996?

7. Who became president when the civil war ended in 1997?

8. Where is Lake Tanganyika located?

9. What is special about Lake Tanganyika?

Where Am I?   Read each sentence. Then look at the words in dark print for the name of the place in each sentence. Write the name of the correct place on your paper.

**Lake Tanganyika   Congo Basin   Kinshasa
Rwanda   Inga Dam**

1. "I am in the area covered by the largest rain forest in Africa."

2. "I am in the capital city of Congo."

3. "I am at the world's longest lake."

4. "I am at a dam on the Congo River."

5. "I am in a country that is Congo's eastern neighbor."

◆ Think and Apply

Sequencing   Number your paper from 1 to 5. Write the sentences on your paper to show the correct order.

In 1994 Hutus in Rwanda killed thousands of Tutsis.

The rebels led by Laurent Kabila won the civil war.

Joseph Mobutu became the leader of Congo.

Fighting between Congo's Hutus and Tutsis led to civil war in Congo in 1996.

The Tutsis won control of Rwanda's government, so Hutu refugees went to Congo.

◆ Journal Writing

Write a paragraph in your journal that tells why Congo has a low standard of living even though it is rich in resources.

CHAPTER 28

The Republic of South Africa

Where Can You Find?
Where can you find a mountain that has a flat top like a table?

Think About As You Read

1. What are South Africa's three landforms?
2. How is South Africa different from other African countries?
3. How has South Africa changed since 1990?

New Words

- ◆ cape
- ◆ escarpments
- ◆ apartheid
- ◆ racial groups
- ◆ minority
- ◆ coloreds

People and Places

- ◆ Cape Agulhas
- ◆ Cape Town
- ◆ Cape of Good Hope
- ◆ Table Mountain
- ◆ Lesotho
- ◆ Orange River
- ◆ Dutch
- ◆ Cape Colony
- ◆ Nelson Mandela
- ◆ F.W. de Klerk

The Republic of South Africa is very different from other African countries. It is not in the tropics. Fewer than half of the people are farmers. It is a rich industrial country with many kinds of factories. South Africa is the only developed country in Africa.

Geography, Climate, and Cities

South Africa is located on the southern tip of the continent. The place that is farthest south is named Cape Agulhas. A **cape** is a point of land that sticks out into a large body of water. The Atlantic Ocean is to the west of this cape. The Indian Ocean is to the east. South Africa has a long coast with beautiful beaches. There are large ports on the coast. These ports help South Africa trade with other countries. Cape Town is a large port city on the Atlantic Ocean side of the Cape of Good Hope. Table Mountain is a famous place in Cape Town. The top of this mountain is flat like a table.

Lesotho is surrounded by South Africa. South Africa has three capital cities. Each one has a different part of the government. What are South Africa's capital cities?

MAP KEY
• City
⊛ Capital city
–·– Border between countries
⌇ River
▲ Mountain peak

ZIMBABWE
MOZAMBIQUE
BOTSWANA
SWAZILAND
KALAHARI DESERT
Pretoria ⊛
Johannesburg •
NAMIBIA
Orange River
Kimberley •
Maseru ⊛
Bloemfontein ⊛
LESOTHO
Durban •
NAMIB DESERT
DRAKENSBERG
ATLANTIC OCEAN
INDIAN OCEAN
GREAT KARROO
East London •
CAPE OF GOOD HOPE
Cape Town ⊛
▲ Table Mountain
Port Elizabeth •
CAPE AGULHAS
NAMIBIA
25°S
30°S
30°S
25°E
30°E

South Africa's flag

Escarpments in South Africa

South Africa has three main landforms. It has a narrow coastal plain. North of the coastal plain are mountains and **escarpments**. Escarpments are steep cliffs. The highest mountains in this region are in Lesotho. Lesotho is a small country surrounded by the country of South Africa. The third landform is a large plateau north of the mountains. This plateau covers most of South Africa. The plateau is covered with grasslands.

The Orange River is the longest river in South Africa. It flows into the Atlantic Ocean. There are several other long rivers. South African rivers have many waterfalls. So the rivers cannot be used for shipping and transportation.

Most of South Africa has a warm, sunny climate. There is not much rain. The climate becomes drier as you travel west. There are deserts in the west and in the northwest.

South Africa is south of the Equator. So it is in the Southern Hemisphere. In the Southern Hemisphere the seasons are the opposite of those in the Northern Hemisphere. In South Africa winter is in June and July.

At least half of South Africa's people live in modern cities. These cities have many industries. They have good roads and public transportation.

History, People, and Apartheid

The first people to live in South Africa were black Africans. Then in the 1600s, the Dutch came to southern Africa from Europe. They settled in the region and called it Cape Colony. Later people from Great Britain settled in the region. During the early 1900s, the British won control of the region and changed its name to South Africa. In 1931 South Africa became an independent country. White South Africans won control of the country's government.

In 1948 the South African government passed **apartheid** laws. These laws allowed white people to control the country's government, natural resources, and money. The government divided the country's people into four **racial groups**. The circle graph on this page shows these groups. Whites are a **minority**, a small part of the population. **Coloreds** are people who have black and white or black and Asian parents. The laws kept the racial groups apart. Each group had its own schools and hospitals. Under apartheid only whites could vote.

Black South Africans and many whites worked to end apartheid laws. One black leader, Nelson Mandela, spent many years in jail because of his work to end apartheid. Finally in 1990 South Africa's president, F.W. de Klerk, began ending these laws. In 1991 all of the apartheid laws ended.

In 1994 South Africa held free elections. For the first time, people from all racial groups could vote. Some people waited in line for many hours to vote. Nelson Mandela became the country's first black president.

Today all South Africans have equal rights. Whites no longer control the government. But white people still have a much higher standard of living.

Resources and Economy

South Africa is very rich in natural resources. The country earns most of its money from exporting

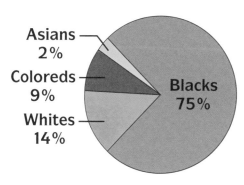

Racial Groups in South Africa

Asians 2%
Coloreds 9%
Whites 14%
Blacks 75%

F.W. de Klerk

Many people in South Africa still have a low standard of living.

New neighborhoods in South Africa give people better places to live.

minerals. Most of the world's gold and diamonds come from South Africa. This country also has coal, iron, copper, silver, and other metals. It has fish from its oceans. There are many kinds of wildlife. But South Africa must buy oil from other countries.

South Africa has more factories than any other African country. South Africans make most of the products they need in their factories. The country imports technology from Western Europe, Asia, and the United States.

Only about one third of all South Africans work at agriculture. Many people also raise sheep and cattle. The country grows most of the food it needs. It exports sugar, fruit, and wine. It also exports lots of wool.

There are big differences between white and black farmers. White farmers have large farms and modern farm machines. They grow cash crops. Most blacks are subsistence farmers. They grow their crops on small farms. They do not have modern farm machines.

South Africa earns more than one billion dollars each year from tourism. Many people enjoy visiting the country's huge national parks. There they can see elephants, lions, monkeys, and other types of wildlife.

Working for a Better South Africa

One of South Africa's biggest problems is that blacks, coloreds, and Asians have a much lower standard of living than white people. Their standard of living is lower because apartheid laws allowed whites to control South Africa's money and resources for many years.

Today almost half of all blacks do not have jobs. Many blacks are homeless. Others live in city slums or small village huts. Malnutrition is a problem because many people do not have enough food.

South Africans can solve these problems by working together. They have already changed their country into a democracy where all people have equal rights. Now they must find ways to raise the standard of living.

Table Mountain in Cape Town

Chapter Main Ideas

1. South Africa is Africa's only developed country.
2. South Africa earns most of its money by exporting gold, diamonds, and other minerals. It also exports food, wool, and wine.
3. In 1994 all racial groups voted in the first free elections. Nelson Mandela became South Africa's president.

BIOGRAPHY

Nelson Mandela (Born 1918)

Nelson Mandela worked to end South Africa's apartheid laws. Mandela is a lawyer who became a leader of the African National Congress. This group worked for equal rights for all racial groups. Mandela was arrested for his work to end apartheid. He spent 27 years in prison. During that time his wife, Winnie, and other people in many nations worked to help him win freedom. He became one of the world's most famous prisoners. At last in 1990, President F.W. de Klerk allowed Mandela to be free.

After leaving prison, Mandela worked with President de Klerk to end apartheid. In 1993 both men received the Nobel Peace Prize for their work. After South Africa's first free elections in 1994, Mandela became the country's first black president.

Journal Writing
Write a paragraph in your journal about Nelson Mandela. Tell how he helped South Africa.

◆ Vocabulary

Finish Up   Choose the word or words in dark print that best complete each sentence. Write the word or words on your paper.

coloreds   **racial groups**   **escarpments**
cape   **minority**   **apartheid**

1. A _____ is a point of land that sticks into a large body of water.

2. Steep cliffs are called _____.

3. Blacks, whites, and Asians are three of South Africa's _____.

4. A group of people that is less than half of the population is a _____.

5. People in South Africa with white and black parents or Asian and black parents are _____.

6. South African laws that kept racial groups apart were called _____ laws.

◆ Read and Remember

Write the Answer   Write one or more sentences to answer each question.

1. What kind of climate does South Africa have?

2. What country does South Africa surround?

3. How do climates differ in the Northern and Southern Hemispheres?

4. What did the apartheid laws do?

5. What event happened in South Africa in 1994?

6. How does South Africa earn money?

7. Why do tourists visit South Africa?

8. How did F.W. de Klerk and Nelson Mandela help South Africa?

9. How is South Africa different from other African countries?

10. What are the three main landforms of South Africa?

11. What are the differences between white and black farmers in South Africa?

12. Why is the standard of living lower for blacks, coloreds, and Asians than it is for whites in South Africa?

13. What place is farthest south on the continent of Africa?

14. What two groups of Europeans settled in South Africa?

15. When did South Africa become an independent country?

◆ Think and Apply

Categories Read the words in each group. Decide how they are alike. Find the best title for each group from the words in dark print. Write the title on your paper.

People Who Settled in South Africa   **South African Resources**
Problems of Black South Africans   **South African Landforms**
Changes in South Africa   **Southern Hemisphere**

1. escarpments
   plateau
   coastal plain

2. gold and diamonds
   coal and iron
   fish and wildlife

3. the end of apartheid laws
   all racial groups could vote
   Nelson Mandela became president

4. black Africans
   Dutch
   British

5. malnutrition
   low standard of living
   subsistence farming

6. winter in June and July
   summer in December and January
   south of the Equator

◆ Journal Writing

South Africa has had a very interesting history. In 1948 the apartheid laws were passed. The people were divided into racial groups. But many changes began in 1990. Write a paragraph in your journal that tells how South Africa has changed since 1990. Tell about three or more changes.

Reading a Population Map

A **population map** shows where people in a region live. It also can show which areas are densely populated and which areas have a lower population density. It can also show the population density of cities. The map key helps you learn the population of a region.

Look at the population map of South Africa. Number your paper from 1 to 7. Then finish each sentence in Group A with an answer from Group B. Write the letter of each correct answer on your paper.

Group A

1. _____ is a port with fewer than one million people.

2. A southeastern port with more than one million people is _____.

3. There are _____ cities in South Africa with more than 5 million people.

4. In the region around Lesotho, there are _____ people per square mile.

5. The deserts have _____ people per square mile.

6. _____ is a city in central South Africa with fewer than one million people.

7. We can conclude from the map that there is greater population density in the _____.

Group B

A. no

B. 25–100

C. east

D. Kimberly

E. Durban

F. Port Elizabeth

G. 0–25

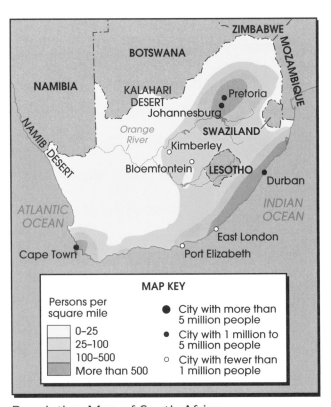

Population Map of South Africa

CHAPTER 29

Working for a Better Africa

Think About As You Read

1. What are some serious problems in Africa today?
2. What are the causes of hunger in Africa?
3. How are Africans trying to solve their problems?

New Words

- ◆ hunger
- ◆ famine
- ◆ desertification
- ◆ overgraze
- ◆ erosion
- ◆ disease
- ◆ tsetse fly
- ◆ AIDS
- ◆ foreign aid
- ◆ experts
- ◆ loans
- ◆ United Nations
- ◆ Peace Corps
- ◆ gorge

People and Places

- ◆ Ethiopia
- ◆ Somalia
- ◆ Victoria Falls
- ◆ Zambia
- ◆ Zimbabwe

Snow-covered mountains, tall waterfalls, sandy beaches, and interesting wildlife make Africa a beautiful continent. It is also rich in natural resources. But Africa is also the poorest region in the world. Africans must solve five big problems so that people can have a higher standard of living.

Rapid Population Growth and Hunger

Africans must solve the problems of **hunger** and rapid population growth. The population is growing faster in Africa than in any other part of the world. African governments cannot provide enough food, jobs, and services for all their people. To solve this problem, people must learn to have smaller families.

Hunger is Africa's second problem. Africa does not grow enough food for its people. Africans grew less food in the 1990s than they did in the 1970s. Hunger leads to malnutrition and poor health.

Many people in Africa have starved during famines.

There are many causes of hunger. **Famine** is one cause. A famine is a terrible shortage of food for a long period of time. Droughts, long periods without rain, often cause famine. Africa has had many droughts. Droughts have caused many famines in Ethiopia. During these famines, millions of people have starved. Developed nations usually send food to help countries that have famines.

A second cause of hunger is that many farmers use traditional methods of farming. They do not have modern machines, good seeds, or good fertilizers. A third cause is that too many farmers grow cash crops such as coffee and cotton. Not enough farmers grow crops Africans can eat.

Many people in Africa continue to use traditional methods of farming.

The fourth cause of hunger is civil wars. Soldiers burn or destroy the crops of their enemies. This happened during the 1992 civil war in Somalia in east Africa. A fifth cause is poor transportation. There are not enough roads, trucks, or trains, so farmers cannot send their crops to the people in cities and villages.

Desertification

Desertification is a third problem in Africa. Desertification means grasslands become smaller and deserts grow larger. Africa's huge Sahara is spreading into the Sahel. So there is less land for growing food. Desertification happens when farmers allow their sheep and cattle to **overgraze**. This means sheep and cattle eat all the grasses and their roots. Then there is no grass to hold down the soil. Without grass the land turns into desert.

Deforestation also causes deserts to grow larger. Deforestation means forests in a region are destroyed. This happens when people chop down too many trees. Without trees there is **erosion**. This means soil is blown away by wind or is washed away by rain. Then the land turns into desert.

Africans are working to stop desertification. In Kenya and in Ethiopia, people are planting many new trees. The new trees prevent erosion. They stop the land from becoming desert.

Illiteracy and Disease

Many people in Africa cannot read a newspaper. Illiteracy, not knowing how to read or write, is Africa's fourth problem. For example, in Ethiopia only one third of the people can read and write. In South Africa many black adults do not know how to read. To solve this problem, governments are building new schools and training more teachers.

When these trees are larger, they will be planted in a forest in Kenya.

Cattle and other animals have overgrazed some land in Africa.

A tsetse fly

Disease, or sickness, is the fifth problem. Insects spread many kinds of diseases throughout Africa. One serious disease is the "sleeping sickness." It is spread by the **tsetse fly**. Africans also become sick from drinking water that is contaminated. **AIDS** is the most serious disease in Africa today. AIDS is spreading fast, and people die from the disease. There is no cure for AIDS yet. But people all over the world are looking for a cure and for better ways to protect people from disease.

Working for a Better Africa

Africans are working to solve their problems. To solve the hunger problem, people are trying to grow more food. In some places people are learning to use modern farming methods and better seeds, fertilizers, and tools. They are also learning better methods to irrigate dry land.

African governments are trying to start new industries and use modern technology. New factories are being built in Africa's cities. Then countries have more products that they can export to earn money. They do not have to depend on one or two cash crops or minerals to earn money.

Foreign aid is helping African nations. Foreign aid is money and help that developed nations give to developing counties. African nations use foreign aid to build dams, roads, factories, and schools. Most African countries receive foreign aid. Foreign aid also means sending **experts** to teach Africans better ways to farm and use their resources. Health experts teach people how to prevent and treat diseases.

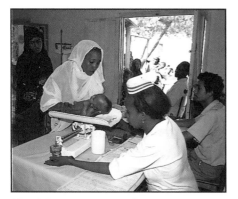
Health experts in Somalia

Sometimes foreign aid is money that one nation lends another. Then African nations must repay the **loans**. Poor countries find it very hard to repay loans. Many countries now have debts that they cannot repay.

Sometimes foreign aid is given to African countries during a famine. In 1992 Somalia had a famine and a civil war. More than a million people were starving. Thousands of American soldiers were sent to Somalia. They helped the **United Nations**, or UN, deliver food

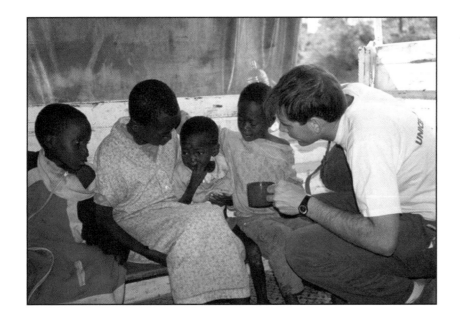

Millions of people starve when there are civil wars.

to starving people. The UN is an organization of countries that work together. The soldiers stayed and helped the people of Somalia for more than a year.

Some Americans are helping Africa by joining the **Peace Corps**. The Peace Corps is an American group that sends its members to help developing countries. In Africa, Peace Corps members start schools. They teach Africans better ways to farm. Other developed countries also send people to Africa to do the same kind of work that Peace Corps members do. They teach Africans how to help themselves.

Africa's people are working to solve their problems. Countries are starting new industries. People are learning better ways to grow food. More children are going to school. People are learning about health care. They are trying to destroy dangerous insects. Africa is slowly becoming a better place to live.

Developed countries send food to countries that have famines.

Chapter Main Ideas

1. Africa's biggest problems are rapid population growth, hunger, desertification, illiteracy, and disease.
2. Millions of Africans have starved during famines.
3. African nations receive foreign aid from the United States and other developed countries.

Victoria Falls

Victoria Falls is one of the world's largest waterfalls. It is more than 300 feet high and more than one mile wide. The falls are located on the border between the countries of Zambia and Zimbabwe. Victoria Falls is formed by the flow of water from the Zambezi River over a high **gorge**. A gorge is a deep valley with steep rocky walls. The river is one mile wide where it flows over the gorge. As the river flows down into the gorge, it sprays huge amounts of water into the air. That spray looks like smoke. It can be seen forty miles from the falls. The sprayed water allows a rain forest to grow near the falls. The falling water sounds like loud thunder. Long ago, Africans called the falls, "smoke that thunders." There is often a rainbow in the sky above the falls.

A dam located near the falls uses waterpower to make electricity for people in the region. A railroad bridge crosses Victoria Falls. That bridge allows people to go across the falls from Zambia to Zimbabwe. People can walk across the bridge. Or they can travel by car or train on the bridge.

Victoria Falls is a wonderful tourist region with many hotels. People who want adventure can go rafting in the gorge below the falls. Other tourists fly over the falls in small airplanes. At night tourists can watch shows with traditional African dancing. All tourists need raincoats at the falls because the spray from the falls makes clothing very wet.

Victoria Falls

Write a sentence to answer each question.

1. **Location**   Where is Victoria Falls?

2. **Human/Environment Interaction**   How does the dam near the falls help people?

3. **Movement**   How can people go across Victoria Falls from Zambia to Zimbabwe?

4. **Place**   What kind of place is Victoria Falls?

5. **Region**   What kind of region is around Victoria Falls?

◆ Vocabulary

Match Up   Finish the sentences in Group A with words from Group B. Write the letter of each correct answer on your paper.

Group A

1. When a region does not have food for a long period of time, there is a _____.

2. When grasslands become deserts, there is _____.

3. When cattle eat all the grass in a region, they have _____.

4. When people get sick, there is _____.

5. The money and help that developed nations give to developing countries is called _____.

Group B

A. foreign aid

B. disease

C. overgrazed

D. desertification

E. famine

◆ Read and Remember

Find the Answer   Find the sentences that tell how people are working to solve Africa's problems. On your paper write the sentences you find. You should find four sentences.

1. In Ethiopia people have planted many trees.

2. Many hungry people have malnutrition.

3. Families have many children.

4. African governments are building schools and training teachers.

5. Africans are building new factories.

6. In some countries many people cannot read or write.

7. Foreign aid money is used to build dams, roads, factories, and schools.

◆ Think and Apply

Drawing Conclusions   Read each pair of sentences. Then look in the box for the conclusion you might make. Write the letter of the conclusion on your paper.

1. The population is growing faster in Africa than in any other region in the world. African governments cannot provide food and services for all the people.

Conclusion: _____

2. African farmers cannot grow enough food because of droughts and civil wars. Farmers grow cash crops like cotton instead of growing food people can eat.

Conclusion: _____

3. Overgrazing kills all the grass of the grasslands, leaving the ground bare. Erosion occurs when forests are destroyed.

Conclusion: _____

4. Only one third of the people of Ethiopia can read. Africa needs more schools and teachers.

Conclusion: _____

5. Many people in Africa have AIDS. Insects spread disease and people get sick from drinking contaminated water.

Conclusion: _____

Conclusions
  A. Illiteracy is a problem in Africa.
  B. African deserts are growing larger because of desertification.
  C. Rapid population growth is a problem in Africa.
  D. There are many causes of hunger.
  E. Disease is a problem in Africa.

◆ Journal Writing

Write a paragraph about two serious problems in Africa today. Tell one or two ways each problem can be solved.

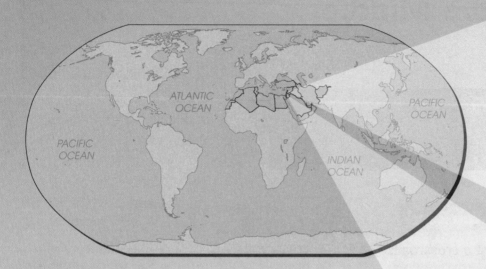

Damascus, Syria

DID YOU KNOW?

▲ **Damascus, the capital of Syria, is the world's oldest capital city.**

▲ **Egypt was one of the first places in the world to have tourists. More than 2,000 years ago, Greeks and Romans came to see Egypt's tombs and temples.**

▲ **The Dead Sea is a saltwater lake that has the saltiest water in the world. The water is almost ten times saltier than regular ocean water.**

▲ **Turkey was the first Muslim country in the Middle East to have a woman as its prime minister.**

The Dead Sea

WRITE A TRAVELOGUE

The Middle East and North Africa have deserts, oil fields, cities, and villages. Look at the photographs of the region in this unit. Choose two different types of places you would like to visit. Write a paragraph in your travelogue that tells why those two places interest you. After reading the unit, choose two other places in different countries that you might want to visit. Tell what is special about those places.

THEME: PLACE

CHAPTER 30

Looking at the Middle East

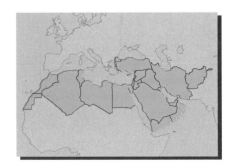

Think About As You Read

1. Why is the Middle East a crossroads region?
2. What resource is very scarce in the Middle East?
3. What religions began in the Middle East?

New Words

- crossroads
- oasis
- steppe climate
- monotheism
- Judaism
- Christianity
- Koran
- Hebrew
- phosphates

People and Places

- Bedouins
- Middle East
- Persian Gulf
- Saudi Arabia
- Riyadh
- Tigris River
- Euphrates River
- Israel
- Jesus
- Muhammad
- Iran
- Lebanon

It is late at night in the desert. A tired young man rests in his tent. He rode a camel for many hours in the hot desert sun. He worked hard taking care of his family's sheep and goats. Tomorrow the man and his family will move again. They are Bedouins. About one million Bedouins live in the deserts of the Middle East. They are nomads who move from place to place in the desert.

The Region and Its Geography

The Middle East is a region that includes countries in southwestern Asia and northern Africa. About 350 million people live in this part of the world. Sometimes people use the name Middle East just for the countries in southwestern Asia. As you read this unit, you will find that some countries in northern Africa share the same culture as the countries in southwestern Asia.

The Middle East is often called a **crossroads** region. It is a crossroads because people often pass

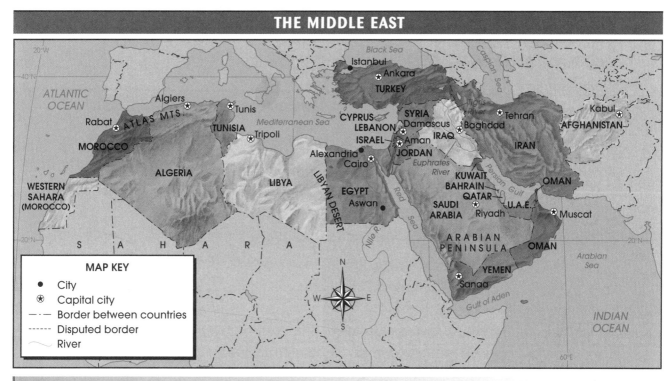

The Middle East is a region that includes countries in southwestern Asia and northern Africa. Which countries are in northern Africa?

through the Middle East to travel between Europe, Asia, and Africa.

Three important bodies of water are found in the Middle East. They are the Mediterranean Sea, the Persian Gulf, and the Red Sea. These seas are used for shipping, trade, and transportation from one continent to another.

There are several different kinds of landforms in the Middle East. There are lowland plains near the coasts. Hills, mountains, and valleys cover many areas away from the coasts. Plateaus also cover parts of the Middle East.

Climates and Rivers

Water is scarce everywhere in the Middle East. It is one of the driest regions in the world. Deserts cover more than half of the region. The huge Sahara Desert covers most of North Africa. A large desert covers most of Saudi Arabia. There are deserts in the northeastern part of the region, too.

A camel market

Because there is water at oases, people can live and farm in the desert.

While traveling through a desert, you might suddenly come to a place with grass and trees. You would be at an **oasis**. An oasis is a place in the desert that has underground water. Many people in the Middle East live near oases and raise sheep, goats, and crops. Riyadh, the capital of Saudi Arabia, was built at a large oasis.

The Middle East has three climates. Most of the region has a desert climate. Areas that are near the Mediterranean Sea have short, rainy winters and long, dry summers. They enjoy a Mediterranean climate. Other areas have a **steppe climate**. A steppe climate has small amounts of rain during the year. Grasses grow in a steppe climate.

Because there is so little rain, millions of people in the region live near rivers. The Nile, the Tigris, and the Euphrates rivers have been important for more than 5,000 years. The soil near these rivers is fertile. People can farm because they use river water to irrigate their fields. Thousands of years ago, people built the world's first cities near these rivers.

Religion and People of the Middle East

Three of the world's important religions began in the Middle East. The three religions teach **monotheism**, a belief in one God. **Judaism**, the religion of the Jews, is the oldest of the three religions. It began in Israel thousands of years ago.

Riyadh, Saudi Arabia

Christianity, the Christian religion, developed from Judaism. This religion follows the teachings of Jesus. Jesus was a Jew who lived about 2,000 years ago. Christians believe that Jesus was the son of God.

Islam began in Saudi Arabia in the year 622. It was started by an Arab leader named Muhammad. He taught that there is one God. Muslims believe that Muhammad was God's messenger. Muhammad's teachings are in the **Koran**. This book is holy to Muslims. The language of the Koran is Arabic. You will learn more about Islam in Chapter 34.

Who are the people of the Middle East? Most people are Arabs. They speak Arabic. In Turkey and Iran the people are Muslims, but they are not Arabs. Today 90 percent of the people in the Middle East are Muslims. Some countries, such as Lebanon and Israel, have a small Christian population. In Israel most people are Jews. Their language is **Hebrew**.

A page from the Koran

Resources and Earning a Living

Oil is the region's most important natural resource. There is more oil in this region than in any other part of the world. Many developed nations do not have enough oil. So they buy oil from the Middle East. Oil-rich countries, such as Saudi Arabia and Iran, earn most of their money by exporting oil.

Studying the Koran

In Israel, road signs are written in Hebrew, Arabic, and English.

These phosphates will be exported to countries around the world.

This region has a few other resources. Some countries have natural gas. Some have iron ore. A few countries have **phosphates**. Phosphates are used in fertilizers.

More than half of the people in the Middle East earn a living by farming. The region has millions of farmers, but most of the land is too dry for farming. Most farmers in the Middle East do not use modern machines and methods. They work the same way farmers worked hundreds of years ago. Israel is one of the few countries in the region where farmers use modern methods.

Most nations in the Middle East are developing countries. Some countries are using the money they earn from selling oil to develop new industries. The standard of living in most of this region is much lower than the standard of living in Western Europe or the United States. Only a small part of the population does factory work.

Most people in the Middle East are poor. Most countries do not grow enough food to feed their people. Wars have been a problem in this region for thousands of years. Today this region still does not have real peace. In the next chapters, find out how people live in four countries of this region.

Chapter Main Ideas

1. The Middle East is the driest region in the world. Most people live near seas or rivers.
2. Judaism, Christianity, and Islam began in the Middle East.
3. Arabs are the largest ethnic group in the Middle East. Most people in this region believe in Islam.

◆ Vocabulary

Finish the Paragraph　Number your paper from 1 to 8. Use the words in dark print to finish the paragraph below. On your paper write the words you choose.

monotheism	Koran	Christianity	steppe climate
crossroads	oasis	Judaism	phosphates

The Middle East is a ___1___ region because people often pass through it to travel between Europe, Asia, and Africa. The Middle East is a dry region. Some places have a ___2___ because they get small amounts of rain during the year. In the desert, water can be found at an ___3___. Three religions of the Middle East teach people to believe in one God. This belief is called ___4___. The religion of the Jews is ___5___. People who follow the teachings of Jesus believe in ___6___. The teachings of Islam are in a book called the ___7___. The region has ___8___, a natural resource used to make fertilizers.

◆ Read and Remember

Write the Answer　Write one or more sentences to answer each question.

1. Where is the Middle East?

2. What three bodies of water are in the Middle East?

3. What are three types of climates in the Middle East?

4. What are three important rivers in the Middle East?

5. What three religions began in the Middle East?

6. What are some of the teachings of Islam?

7. What is the most important resource in the Middle East?

8. Why is only a small amount of land in the Middle East good for farming?

9. What religion do most people in the Middle East follow?

10. What capital city was built at an oasis in the desert?

◆ Think and Apply

Fact or Opinion Number your paper from 1 to 10. Write **F** on your paper for each fact. Write **O** for each opinion. You should find four sentences that are opinions.

1. The Mediterranean Sea is to the north of North Africa.

2. More people should live near the Nile River.

3. People in the Middle East use river water to irrigate their farms.

4. It is better to live near an oasis than to live near a river.

5. Monotheism began in the Middle East.

6. Most people in the Middle East believe in Islam.

7. The people of the Middle East should export more phosphates.

8. Nations that have oil should lower the price of oil.

9. The Middle East is one of the driest regions in the world.

10. Most nations in the Middle East are developing countries.

◆ Journal Writing

You have been reading about the Middle East. You learned that it is a region in southwestern Asia and northern Africa. You learned what the land and climate are like. You also learned something about the people of the region. Imagine living in the Middle East. Would you want to be a Bedouin, a farmer, or a city worker? Write a paragraph in your journal that tells which way of life you would want and why.

CHAPTER 31

Egypt: The Gift of the Nile

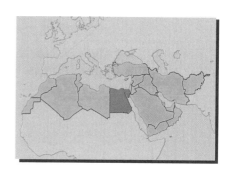

Where Can You Find?
Where can you find the only subway system in Africa?

Think About As You Read

1. How are the nations of North Africa alike?
2. How does the Nile River help Egypt?
3. How has the Aswan High Dam helped and hurt Egypt?

New Words

◆ Islamic fundamentalism
◆ deposited
◆ silt
◆ chemical fertilizers
◆ ancient
◆ pharaohs
◆ pyramids
◆ manganese
◆ unfavorable balance of trade

People and Places

◆ Morocco
◆ Algeria
◆ Tunisia
◆ Libya
◆ Sinai Peninsula
◆ Suez Canal
◆ Nile Delta
◆ Nile Valley
◆ Egyptians
◆ Aswan High Dam
◆ Lake Nasser
◆ Alexandria

For 5,000 years the people of Egypt have lived on land near the Nile River. The Nile River is just as important to Egypt today as it was long ago.

The Region of North Africa

Egypt has the largest population in North Africa. Morocco, Algeria, Tunisia, and Libya are other nations in the region. The Sahara Desert covers most of these countries. Most people in this region live near the Mediterranean Sea. People in Egypt also live along the Nile River. The Nile is the only long river in North Africa. North Africans use the Mediterranean Sea to trade with Europe and other parts of the Middle East.

All North African countries are developing nations. Most people work as subsistence farmers. Algeria and Libya have rich oil resources. These countries are using the money they earn from selling oil to become more developed.

The only long river in North Africa is in Egypt. What is the name of the river?

EGYPT

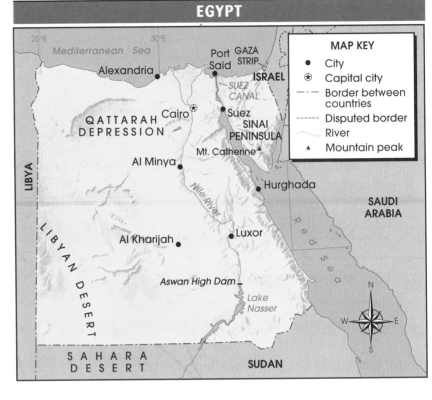

MAP KEY

- City
- ⊛ Capital city
- –·– Border between countries
- ----- Disputed border
- ～ River
- ▲ Mountain peak

Egypt's flag

The Suez Canal

In all of the countries in North Africa, most people are Arabs. The official language is Arabic. Their religion is Islam.

A movement called **Islamic fundamentalism** is changing North Africa and other countries in the Middle East. This means many Muslims have turned away from western values. Instead they follow the strictest rules of Islam. Because of Islamic fundamentalism, more people are wearing traditional Muslim clothing. Sometimes the leaders of this movement force people to follow the ways of Islam. When this happens there is less freedom.

Egypt: Climate, Landforms, and the Suez Canal

Egypt is in the northeastern part of North Africa. The country has a hot, dry climate. Deserts cover most land west and east of the Nile River. Deserts also cover the Sinai Peninsula. The Sinai Peninsula is part of eastern Egypt. Rocky hills are part of the Sinai and the eastern desert. A large plateau covers most of western Egypt.

Egypt earns large amounts of money from ships that use the Suez Canal. The canal allows ships to sail between the Mediterranean Sea and the Red Sea. Ships from many nations use the Suez Canal. The canal makes it faster to sail from Europe to Asia. Ships pay money to Egypt to use the canal.

The Nile River and the Aswan High Dam

The Nile is Egypt's only river. It is more than 4,000 miles long. It begins far from Egypt in central Africa near the Equator. The Nile flows from south to north. In the north, at the mouth of the river, the Nile empties into the Mediterranean Sea. For thousands of years, the river has **deposited** soil at the mouth of the river. This soil built up and formed a delta. The Nile Delta has very fertile soil.

The Nile Delta seen from outer space

Egypt is called the "Gift of the Nile." It has this name because the waters of the Nile and the fertile soil have been used to grow crops on dry desert land. The land around the Nile is called the Nile Valley. Every summer, for thousands of years, the river flooded the Nile Valley. The flood waters left **silt** on the land around the river. Silt is tiny pieces of soil. The silt made the Nile Valley very fertile. For thousands of years, people have farmed the Nile Valley. For thousands of years, people have used Nile water to irrigate farms in the Nile Valley.

The Nile Valley and the Nile Delta have less than five percent of Egypt's land. But most of Egypt's people live on this land. Most of Egypt's cities are in the Nile Valley.

Between 1960 and 1968, Egyptians built a huge dam on the Nile River. It is called the Aswan High Dam. The

Silt from the Nile River made the Nile Valley very fertile.

The water from Lake Nasser irrigates fields in the Nile Valley all year long.

The Aswan High Dam

dam stopped the floods of the Nile River. It has helped Egypt in several ways. The dam saves the Nile's water in a large lake, or reservoir, called Lake Nasser. Water from the reservoir is used to irrigate fields all year long. Egypt now has much more farmland. Egyptians grow twice as much food as they did before 1968. The dam also uses waterpower to make large amounts of electricity.

But the Aswan High Dam has also caused four problems. First, the Nile Valley no longer has floods from the river. So the Nile Valley no longer gets new silt from the river. The soil is less fertile. Farmers must use **chemical fertilizers** on the soil. Many poor farmers do not have enough money to buy these fertilizers. Second, chemical fertilizers have polluted the Mediterranean Sea. The fertilizers have killed some of the fish in the sea. Third, erosion is a problem at the Nile Delta. The delta no longer gets new soil each year. The waves of the Mediterranean Sea are washing away land at the mouth of the river. Fourth, tiny snails that carry disease now live all year in Lake Nasser. People who swim, bathe, or wash clothes in the lake water can become very sick.

Egypt's History, People, and Cities

Egypt's history began more than 5,000 years ago. People farmed the Nile Valley. They built cities near the Nile. Egyptians traded with countries around the Mediterranean Sea.

Ancient Egypt was ruled by kings called **pharaohs**. Some of the pharaohs forced slaves to build huge **pyramids** in the desert. The pharaohs were buried in these pyramids when they died. Many pyramids are still standing today.

One of Egypt's pyramids

Today Egypt is the largest Arab country in the Middle East. About 60 million people now live in Egypt. It is one of the leaders in the Arab world. More Arabic books and movies are made in Egypt than in any other country.

Egypt has a population that is growing very fast. About half of the people live in cities. Cairo is Egypt's largest city. It is on the Nile River in northern Egypt. Cairo is more than 1,000 years old. About 10 million people live in and around the city. There are modern hotels, homes, and office buildings. Cairo has the only subway system in all of Africa. But Cairo has millions of poor people. They live in tiny apartments. They do not have electricity or running water in their homes.

Cairo is on the Nile River.

Alexandria is Egypt's second largest city. It is a large port on the Mediterranean Sea.

Resources and Economy

Egypt's most important resource is oil. Egypt exports its oil to Europe and the United States. It also has iron ore, **manganese**, and coal.

Cutting sugarcane

Almost half of the people earn a living by farming. Cotton is the main cash crop. Egypt also exports large amounts of dates. Farmers grow sugarcane, rice, wheat, and oranges. But they do not use modern methods. Egypt must import half of the food it needs.

Egypt earns more than a billion dollars each year from tourism. Tourists visit Egypt's pyramids and museums. They also enjoy boat trips on the Nile.

Some Egyptians do factory work. Egyptian factories make cloth, food products, and other goods. Egypt has an **unfavorable balance of trade**. This means Egypt buys more goods from other countries than it exports. The United States has given Egypt billions of dollars in foreign aid.

Today Egypt has a low standard of living. Cities are very crowded. Millions of people are very poor. The country needs more farmland and more factories. Egyptians are trying to solve these problems. They want their country to continue to be a great leader in the Arab world.

A street in Alexandria

Chapter Main Ideas

1. The countries of North Africa are developing nations. Most people are Arabs and believe in Islam.
2. Egypt is the "Gift of the Nile." The Nile allows people to grow food in the desert near the river.
3. The Aswan High Dam has given Egypt farmland and electricity. But the dam has caused erosion of the Nile Delta and has harmed fish in the Mediterranean Sea.

◆ Vocabulary

Finish Up   Choose the word or words in dark print that best complete each sentence. Write the word or words on your paper.

pharaohs	**pyramids**	**unfavorable balance of trade**	**silt**
chemical fertilizers	**ancient**	**Islamic fundamentalism**	

1. A return to the strict rules of Islam is _____.

2. The floods of the Nile River left soil called _____ on the land around the river.

3. The history of _____ Egypt began 5,000 years ago.

4. Long ago the kings of Egypt were called _____.

5. Thousands of years ago, Egyptian kings were buried in _____.

6. When a nation imports more than it exports, it has an _____.

7. Farmers in Egypt use _____ to improve their soil.

◆ Read and Remember

Where Am I?   Read each sentence. Then look at the words in dark print for the name of the place for each sentence. Write the name of the correct place on your paper.

Alexandria   **Nile Valley**   **Lake Nasser**   **Sinai**   **Cairo**   **Nile Delta**

1. "I am on a peninsula in eastern Egypt."

2. "I am on land formed by the build-up of soil at the mouth of the Nile River."

3. "I am on the fertile land around the Nile River."

4. "I am at a huge lake made by the Aswan High Dam."

5. "I am in the largest city on the Nile River."

6. "I am at a large port on the Mediterranean Sea."

Write the Answer Write one or more sentences to answer each question.

1. Why is Egypt the "Gift of the Nile"?

2. How has the Aswan High Dam helped Egypt?

3. How has the Aswan High Dam hurt Egypt?

4. What are Egypt's resources?

5. What is the climate of Egypt?

6. Which waterway earns Egypt large amounts of money?

7. What two bodies of water does the Suez Canal connect?

8. Where do most of Egypt's people live?

9. What kind of place is Cairo?

10. How has Islamic fundamentalism changed North Africa and other countries in the Middle East?

◆ Think and Apply

Categories Read the words in each group. Decide how they are alike. Find the best title for each group from the words in dark print. Write the title on your paper.

Suez Canal   **Aswan High Dam**   **North African Nations**   **Nile River**

1. mostly desert
   most people are Muslim Arabs
   the most people live near the
      Mediterranean Sea

2. makes electricity
   holds water in Lake Nasser
   stops floods of the Nile River

3. waterway joins Mediterranean and
      Red seas
   ships sail between Europe and Asia
   Egypt earns money

4. starts near the Equator
   more than 4,000 miles long
   empties into the Mediterranean Sea

◆ Journal Writing

Imagine that you are a tourist in Egypt. Write a paragraph in your journal that tells about three or more places you would want to visit. Tell why you would visit those places.

Reviewing a Resource Map

The **resource map** on this page shows where Egypt's important minerals and crops are found. Use the map key to find out which resources are shown. Then write the word or words that best complete each sentence.

1. A food crop that is found at oases in the desert is _____.

   dates   cotton   rice

2. Two mineral resources in the Sinai Peninsula are _____.

   coal and iron   gold and phosphate

   oil and manganese

3. Two important minerals in northern Egypt are _____.

   coal and gold   oil and natural gas

   manganese and phosphate

4. A food crop near Luxor is _____.

   wheat   corn   rice

5. An important cash crop near Luxor is _____.

   rice   corn   cotton

6. There are _____ food crops in the Sinai Peninsula.

   no   a few   many

7. A food crop that grows in the Nile Delta is _____.

   dates   rice   apples

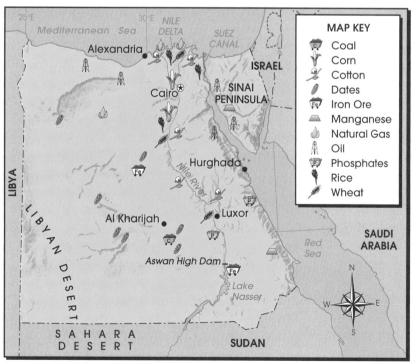

Resource Map of Egypt

CHAPTER 32

Israel: A Jewish Country

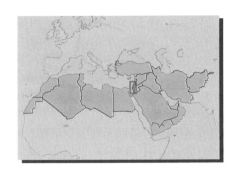

Where Can You Find?
Where can you find the lowest spot on Earth?

Think About As You Read

1. How is Israel different from other countries in the Middle East?
2. How do Israelis grow food in the desert?
3. How did Israel become a Jewish country?

New Words

- scarce
- drip irrigation
- kibbutz
- software
- homeland
- Palestinian Liberation Organization (PLO)

People and Places

- Israelis
- Negev Desert
- Sea of Galilee
- Dead Sea
- Jerusalem
- Tel Aviv
- Beersheba
- Palestinians
- Jordan
- Old City
- Gihon Spring

Where can you find the lowest spot on Earth? Where can you find the only Jewish country in the world? The answer to both questions is Israel.

Israel's People and Landforms

More than 5 million people live in Israel. The people of Israel are Israelis. About 83 percent of Israelis are Jews. Arabs are the rest of the population. The Arabs are Muslims and Christians. The laws of Israel allow freedom of religion. Hebrew and Arabic are the two official languages of Israel.

Israel is a small country at the eastern end of the Mediterranean Sea. There are coastal plains near the sea. These plains have beaches, farms, and cities. The Negev Desert is in the south. Hills cover northern Israel. The Sea of Galilee is a lake in the north. Pipes carry water from this lake to all parts of Israel.

The Dead Sea is a saltwater lake in Israel. It is part of the Great Rift Valley. The Dead Sea is the lowest

ISERAEL

MAP KEY
- • City
- ✪ Capital city
- –·– Border between countries
- ----- Disputed border
- ⌒ River
- ▲ Mountain peak

LEBANON
GOLAN HEIGHTS
SYRIA
Bay of Haifa
Sea of Galilee
Haifa
Netanya
Tel Aviv
WEST BANK
Jordan River
Jerusalem
Dead Sea
Hebron
GAZA STRIP
Gaza
Beersheba
Mediterranean Sea
EGYPT
NEGEV DESERT
JORDAN
Mt. Ramon
Wadi al Arabah
SUEZ CANAL
SINAI
Gulf of Suez
Gulf of Aqaba
Elat
SAUDI ARABIA
34°E
32°N
32°E
N S E W

Deserts cover more than half of the Middle East. What large desert is found in Israel?

Israel's flag

spot on Earth. Its water is so salty that no plants or fish can live in the Dead Sea.

Cities and Farms

Most Israelis live in cities. Jerusalem is the capital and the largest city. It is Israel's center of government and culture. Tel Aviv is a large city on the Mediterranean coast. It is the country's center of business and industry. Beersheba is the largest city in the Negev. The city began thousands of years ago near an oasis.

Only five percent of Israelis are farmers. They use modern machines and lots of fertilizer. They grow most of the food Israel needs. Israel exports food, wine, and flowers to many nations.

Israelis know how to grow food in the desert. They have built pipelines to carry water from the Sea of Galilee to the Negev. Water is **scarce**, so Israelis are careful not to waste it. They use a method called **drip irrigation**. This means hoses carry water to each plant. Each plant gets just enough water to grow. Israeli farmers are teaching people in developing countries better ways to grow food.

Drip irrigation in the Negev

257

A kibbutz on the Sea of Galilee

This fruit grows on a kibbutz.

Many farms in Israel are **kibbutz** farms. A kibbutz is owned by all of its members. Members share the work on the kibbutz. They do not get paid. But the kibbutz gives its members all the things they need. There are schools and doctors. Members eat together in a large dining room. A kibbutz may have a hotel where tourists can stay during a vacation. Only a small part of Israel's people live on kibbutz farms. But these people grow a lot of Israel's food.

Resources, Economy, and Government

Israel has very few resources. There is no coal, oil, waterpower, natural gas, or iron ore. The nation mines salt and minerals from the Dead Sea and the Negev. Israel must import most raw materials that it needs for its industries.

Israel must import oil and coal to produce electricity. But Israel has plenty of sunlight. Almost every Israeli house uses solar energy, or energy from the sun, to make hot water.

Israel is a developed country. All children go to school for many years. Most Israelis work at service jobs. Many service jobs are in the tourist industry. About one fifth of the people have factory jobs. Some Israeli factory products are computer **software**, food products, clothing, and weapons. Israeli diamond cutters make more polished diamonds than any other country. Israel does not earn enough money because it has an unfavorable balance of trade. It imports many more products than it exports.

Israel is a democracy. All citizens, Arabs and Jews, can vote. Citizens vote for members of Parliament. A prime minister leads the country.

Israel's History and Problems

Thousands of years ago, the Jews ruled Israel. They believed God gave the land of Israel to their people. Later the land was conquered by other peoples. Most of the Jews were forced to move to other countries. But some Jews have always lived in Israel. Many Arabs came to live in the region when the Jews left.

During the late 1800s, Jews began to move back. They wanted to make the region a Jewish **homeland** again. During World War II, millions of Jews were killed during the Holocaust. After the war many Jews wanted Israel to be their home again. In 1948 the United Nations formed the country of Israel.

A large group of Arabs left Israel when the new Jewish country was formed. These Arabs are called Palestinians. The Palestinians later wanted the area back. Since 1948 there has been fighting between Israelis and Palestinians over the land.

The people who live in this building use solar energy to make hot water.

Near downtown Tel Aviv

An Israeli soldier

These people from Ethiopia are new immigrants to Israel.

Fighting and wars have been a big problem for Israel. Since 1948 neighboring Arab countries have fought four wars against Israel. Each time Israel remained free. Israel still has some of the land that it took from these Arab countries during a war in 1967.

In 1979 Egypt became the first Arab country to sign a peace treaty with Israel. Israel returned to Egypt some of the land it had captured in 1967. In 1994 Jordan and Israel also signed a peace treaty. Other Arab countries refuse to make peace with Israel.

The Palestinians formed an organization called the **Palestinian Liberation Organization**, or PLO. The Israelis and the PLO have agreed to allow Palestinians to rule themselves in some of the areas that Israel captured in 1967. Some Israelis fear the PLO wants to rule all of Israel. Other people hope the Israelis and the Palestinians will live together peacefully.

Terrorism has been a problem in Israel. Arab terrorists have attacked buses, stores, and schools. A few Jewish terrorists have attacked Arabs. Many people have been killed.

Because of the wars, Israel has a large army. All men and women must serve in the army. Israelis pay very high taxes in order to pay for their army.

Jews from every part of the world have moved to Israel. Since 1980 starving Jews from Ethiopia have escaped to Israel. Thousands of Russian Jews have also made Israel their home. Israelis are working hard to give homes and jobs to the new immigrants.

Every year tourists visit Israel. They visit holy places in Jerusalem. They swim in the Dead Sea. Tourists enjoy this modern country in the Middle East.

Chapter Main Ideas

1. Israel is a developed country. Farmers use modern methods to grow most of the country's food.
2. Israel has few natural resources. It imports raw materials to make factory products.
3. Israel has fought and survived four wars since 1948.

Jerusalem

Jerusalem is a city that is holy to Jews, Christians, and Muslims. Jerusalem is located in central Israel. It has a cooler climate than cities on the coast because of its higher elevation.

Three thousand years ago, King David, the king of Israel, made Jerusalem the capital of his country. Large Jewish temples were built in the city. Walls were built around Jerusalem to keep out enemies. The oldest part of Jerusalem still has walls around it. This area is now called the Old City. Newer areas surround the Old City.

Water has always been scarce in Jerusalem. Long ago, people in Jerusalem got their water from the nearby Gihon Spring. The people were afraid that they might not be allowed to leave the city to get water during a war. So they built an underground tunnel. It went from the Old City to the Gihon Spring. That tunnel saved the people of Jerusalem during many wars.

Today Jerusalem is a region of culture and education. The city is famous for its museums, schools, and large university. Tourists from every part of the world visit its holy places and museums. Some tourists get their feet wet in the ancient tunnel leading to the Gihon Spring. Then they can return to their hotel to swim in a modern pool.

The Old City in Jerusalem

Write a sentence to answer each question.

1. **Location** What is the location of Jerusalem?

2. **Place** Why is Jerusalem a special place?

3. **Movement** Why do tourists come to Jerusalem?

4. **Human/Environment Interaction** Why did people build an underground tunnel to the Gihon Spring?

5. **Region** What kind of region is Jerusalem?

◆ Vocabulary

Match Up Finish the sentences in Group A with words from Group B. On your paper write the letter of each correct answer.

Group A

1. When there is not enough water, water is _____.

2. When hoses bring just enough water to each plant, there is _____.

3. A farm where people share the work and are given everything they need is a _____.

4. Programs that are made for a computer are _____.

5. A country that is home for a group of people is a _____.

6. An organization that wants a country for Palestinians is the _____.

Group B

A. homeland

B. drip irrigation

C. Palestinian Liberation Organization

D. scarce

E. software

F. kibbutz

◆ Read and Remember

Finish Up Choose the word or words in dark print that best complete each sentence. Write the word or words on your paper.

Tel Aviv	**homeland**	**Sea of Galilee**	**terrorists**
imports	**Jordan**	**Dead Sea**	**Jerusalem**

1. Pipes carry water from the _____ to all parts of Israel.

2. The lowest spot on Earth is at the _____.

3. A city on the Mediterranean Sea with lots of businesses and factories is _____.

4. Israel _____ raw materials for its industries.

5. Israel was created to be a Jewish _____ after the Holocaust.

6. Egypt and _____ are the only Arab countries that have signed peace treaties with Israel.

7. Israeli buses, stores, and schools have been attacked by _____.

8. Israel's capital and largest city is _____.

◆ Think and Apply

Finding Relevant Information   Imagine you are telling your friend why Israel is a developed country. Read each sentence below. Decide which sentences are relevant to what you will say. On your paper write the relevant sentences you find. You should find six relevant sentences.

1. Israel uses solar energy to make hot water.

2. Most people work at service jobs.

3. A small group of farmers grow most of the country's food.

4. Israel is a democracy with a Parliament.

5. Israel imports the oil it needs.

6. Israeli factories make computer software.

7. Jewish immigrants from many countries have moved to Israel.

8. Israel gets some minerals from the Dead Sea.

9. Israeli farmers use modern machines and fertilizer.

10. Most people live in cities.

◆ Journal Writing

You have been reading about Israel, one of the countries in the Middle East. Write a paragraph in your journal that tells three ways in which Israel is different from other countries in the Middle East.

Comparing a Climate Map With a Population Map

By comparing a **climate map** with a **population map**, you can learn about a region. The population map on this page shows us that southern Israel has a much lower population density than northern Israel. The climate map shows us that southern Israel has a desert climate. So we can conclude that the south has fewer people because of its desert climate.

Study the map keys on both maps. Then use the words in dark print to finish each sentence. Write the sentences on your paper.

500–1,500   4   Sea of Galilee   Mediterranean   Beersheba   Elat

1. All of the cities near the _____ have fewer than 100,000 people.

2. Israel has a higher population density where there is a _____ climate.

3. The only city with a desert climate that has population of more than 100,000 is _____.

4. There are _____ cities on the Mediterranean Sea that have more than 100,000 people.

5. The region around Herzliya has _____ people per square mile.

6. The southern city of _____ has a desert climate.

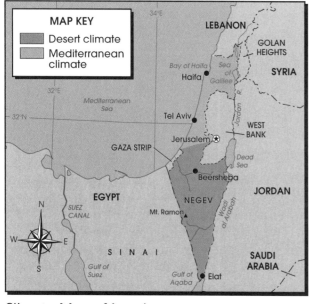

Climate Map of Israel

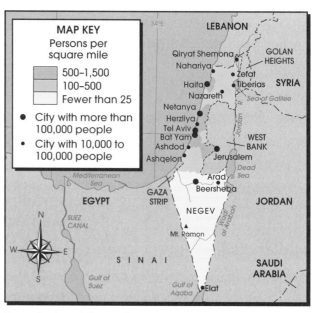

Population Map of Israel

CHAPTER 33

Turkey: A Country on Two Continents

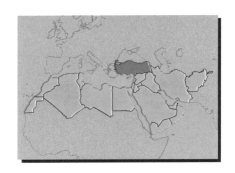

Where Can You Find?

Where can you find a city with bridges that connect Europe with Asia?

Think About As You Read

1. Why does Turkey have ties to both Europe and the Middle East?
2. What are Turkey's landforms and climates?
3. How did Mustafa Kemal Ataturk change Turkey?

New Words

- strait
- Ottoman Empire
- secular
- shish kebab
- chromite

People and Places

- Bosporus Strait
- Thrace
- Anatolia
- Turks
- Ottoman Turks
- Mustafa Kemal Ataturk
- Kurds
- Istanbul
- Constantinople
- Hagia Sofia
- Ankara

Turkey is the only country in the Middle East with land in both Europe and Asia. The people of Turkey have ties to both regions.

Turkey's Landforms, Climate, and Resources

Turkey is a country on two continents. The Bosporus Strait separates the European part of Turkey from the Asian part. A **strait** is a narrow body of water that connects two larger bodies of water. Ships can sail from the Black Sea through the Bosporus Strait. From there they sail into the Mediterranean Sea.

The European part of Turkey has fertile hills and plains. This part of the country is called Thrace. Most of Turkey is in Asia. The Asian part of Turkey is called Anatolia. Anatolia has coastal plains in the north and south. Most of Anatolia is covered with a large dry plateau. High mountains surround the plateau.

People enjoy a Mediterranean climate near the Black and Mediterranean seas. Turkey's mountains

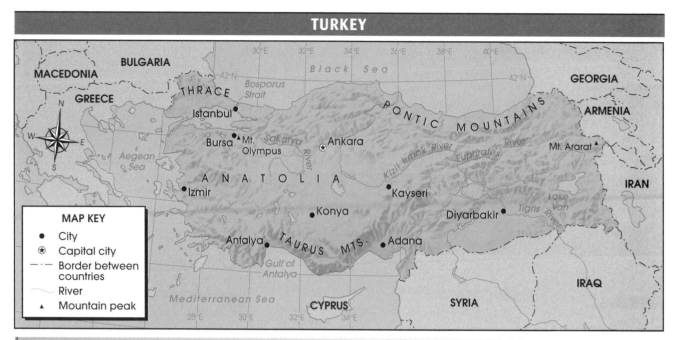

Anatolia is the eastern part of Turkey. Most of Anatolia is covered with a large plateau. What two mountain chains surround the plateau?

Turkey's flag

Mustafa Kemal Ataturk

and plateaus in Asia have a steppe climate. They get some rain and snow. Summers are hot and winters are very cold.

Turkey has two of the important rivers in the Middle East. The Tigris and Euphrates rivers begin in Turkey.

History, People, and Government

People first settled in Turkey thousands of years ago. Later the ancient Greeks ruled Turkey. Then the Romans ruled Turkey. Greek and Roman ruins can be found in Turkey today.

In 1071 Turks from Asia conquered Turkey. It became a Muslim country. Then in 1453 the Ottoman Turks conquered Turkey. Turkey became part of the large **Ottoman Empire**. The empire fell apart after World War I. Turkey became an independent country in 1922.

In 1923 Mustafa Kemal Ataturk became Turkey's leader. He is called the "Father of Modern Turkey." He changed Turkey into a modern **secular** country. A secular country does not have a religious government. Kemal made the religion of Islam completely separate from the government. All of the people had freedom of religion. Women were allowed to vote. Kemal encouraged people to wear western clothes. All of these changes can be seen in Turkey today.

Today there are about 62 million people in Turkey. The official language is Turkish. About 99 percent of the people in Turkey are Muslims. Small groups of Christians and Jews also live in Turkey. Turkey is the only Muslim country in the Middle East that is a democracy.

About 12 million Kurds live in Turkey. They are the largest minority group in Turkey. The Kurds want their own country. But Turkey does not want the Kurds to rule themselves.

The Turks are proud of their food and culture. People around the world enjoy strong Turkish coffee. **Shish kebab**, small pieces of meat and vegetables on a stick, is a popular Turkish food.

Good storytelling is part of Turkey's culture. Turkish steam baths are also part of the culture. Men and women visit the steam baths at separate times. The baths are places to relax and to talk with friends.

Resources, Cities, and Earning a Living

Turkey has many minerals. It is the only country in the Middle East with large amounts of coal. Turkey also has oil, copper, iron ore, bauxite, and **chromite**. Chromite is a mineral that contains chrome. When mixed with other metals, chrome is used to make car bumpers, door handles, and pans. Turkey has not developed many of its mineral resources.

Coffee seller

A Turkish farming family

This bridge over the Bosporus Strait connects Europe and Asia.

Hagia Sofia

Turkey is a developing country. About half of the people live in cities. Most city people have a higher standard of living than people in villages.

Istanbul is Turkey's largest city and port. It was once called Constantinople. About 10 million people live in and around Istanbul. Part of the city is in Europe. The other part is across the Bosporus Strait in Asia. Two bridges connect the city. Istanbul has many factories. It also has more than 1,000 mosques. One of the most famous places in Istanbul is Hagia Sofia. It was first used as a church. Later it became a mosque. Today Hagia Sofia is a museum.

Ankara is Turkey's capital and second largest city. Many people in Ankara work at service jobs. The city also has many factories.

About 60 percent of the people in Turkey are farmers. Many farmers still use traditional farm methods. Animals are used to do much of the farm work. Still, Turkey grows enough food for its needs.

About 11 percent of the people work at factory jobs. Turkey now earns more money from its factory products than from its farm products. The country also earns billions of dollars from tourism.

Turkey trades with the United States and with countries in Europe and Asia. Turkey imports more products than it exports. So the country has an unfavorable balance of trade.

Looking at the Future

Turkey is working to raise its standard of living. More factories are being built. Most children go to school. Cities have good public transportation.

Turkey also has problems to solve. There are not enough jobs. Many Turks have moved to Germany and other countries in Europe in order to find work. There are not enough schools for children in villages.

Islamic fundamentalism is another problem. Many people do not want Turkey to be a secular country. They want religious leaders to lead Turkey's government. Most Turks want their country to continue to be a secular nation.

Many Turks want stronger ties with Europe. Since 1952 Turkey has been a member of NATO. This organization protects Europe. Turkish soldiers have worked with soldiers from other NATO countries. Now Turkey wants to be a member of the European Union. So far, the European Union has not allowed Turkey to become a member.

Turks are working hard to develop their country. Will their country form closer ties with Muslim countries in the Middle East? Will it become part of the European Union? Today Turkey continues to have ties to both regions.

This young Turkish boy in Anatolia is watching sheep.

Making telephones in a factory in Turkey

Chapter Main Ideas

1. Western Turkey is on the continent of Europe. Eastern Turkey is in Asia.
2. Turkey is a developing country. More than half of the people are farmers.
3. Turkey wants stronger ties to Europe. It belongs to NATO. It wants to join the European Union.

◆ Vocabulary

Find the Meaning   On your paper write the word or words that best complete each sentence.

1. A **strait** is a narrow body of _____ that connects two larger bodies of water.

   land   water   sand

2. **Shish kebab** is a popular type of Turkish _____.

   church   food   steam bath

3. **Chromite** is the _____ from which we get chrome.

   plant   animal   mineral

4. _____ became part of the **Ottoman Empire** in 1453.

   Turkey   Greece   Rome

5. A **secular** government is separated from _____.

   taxes   religion   communism

◆ Read and Remember

Finish the Paragraph   Number your paper from 1 to 10. Use the words in dark print to finish the paragraphs below. Write the words you choose on your paper.

| Mustafa Kemal Ataturk | Kurds | democracy |
| Muslim | women | |

Since 1071 Turkey has been a __1__ country. In 1923 __2__ became the leader of Turkey. He helped Turkey become a modern country. He encouraged people to wear western clothes. Laws were changed to allow __3__ to vote. Also, all of the people were given freedom of religion. Twelve million __4__ are a large minority in Turkey. The government of Turkey is a __5__.

Ankara	European Union	developing	NATO	Istanbul

Turkey is a ___6___ country where more than half of the people work as farmers. One part of the city of ___7___ is in Europe and the other part is in Asia. The capital is the city of ___8___. Turkey has soldiers in the organization called ___9___. Now Turkey wants to join the ___10___.

♦ Think and Apply

Compare and Contrast   Copy the Venn diagram shown below on your paper. Then read each phrase. Decide whether it tells about Thrace, Anatolia, or all of Turkey. If it tells about either part of Turkey, write the number of the phrase in the correct part of the Venn diagram on your paper. If the phrase tells about the entire nation, write its number in the center of the diagram on your paper.

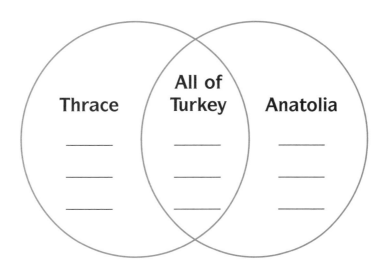

1. steppe climate on plateau and mountains
2. Muslim country
3. women vote in elections
4. small region
5. west of Bosporus Strait
6. east of Bosporus Strait
7. earns money from tourism
8. larger part of Turkey
9. fertile hills and plains

♦ Journal Writing

Imagine telling a friend that Turkey is different from other nations in the Middle East. Write a paragraph in your journal that explains three or more ways Turkey is different from other nations in the region.

Reviewing a Distance Scale

A **distance scale** compares distances on a map with distances in the real world. We use a distance scale to find the distance between two places. One inch on this map represents 300 miles in Turkey.

Look at the map of Turkey on this page. Use your ruler to measure distances. Then use the distance scale to find the answers. On your paper write the word or words that finish each sentence.

1. The distance between Istanbul and Konya is _____.

½ inch   1 inch   2 inches

2. The distance between Istanbul and Ankara is about _____ miles.

50   300   1,000

3. There are about _____ inches between Ankara and Lake Van.

¾   1½   2¼

4. The distance between Ankara and Lake Van is about _____ miles.

200   400   600

5. There are almost 2 inches between Troy and Adana. The distance is almost _____ miles.

50   75   600

6. By using a ruler and a distance scale, we know the distance from Troy to Lake Van is about _____.

90 miles   900 miles   2,900 miles

7. By using a ruler and a distance scale, we know the distance between Ankara and Konya is _____ miles.

10   15   150

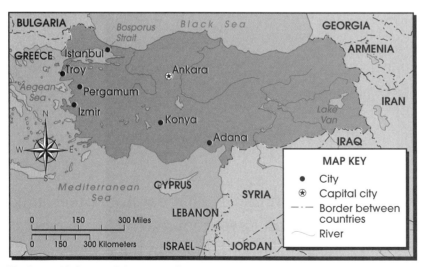

Turkey: Using a Distance Scale

CHAPTER 34

Saudi Arabia: An Oil-Rich Desert Nation

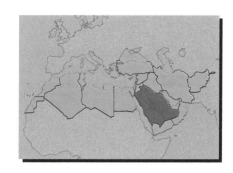

Where Can You Find?
Where can you find sand dunes that are 1,000 feet high?

Think About As You Read

1. What kind of government does Saudi Arabia have?
2. How does Islam affect Saudi Arabia?
3. How have Saudi Arabia's oil resources helped the nation?

New Words

- sand dunes
- foreign
- absolute monarchy
- modest
- aba
- Hajj
- Kaaba
- OPEC
- desalination plants

People and Places

- Arabian Peninsula
- Empty Quarter
- Jidda
- Mecca
- Medina
- King Fahd
- Great Mosque

Saudi Arabia has more oil deposits than any other nation in the world. It is using the money it earns from selling oil to become a modern, developed country.

Landforms, Climate, and Cities

Saudi Arabia covers most of the Arabian Peninsula. There are also a few other small countries on this peninsula. Find Saudi Arabia on the map on page 274.

Saudi Arabia has mountains in the west near the Red Sea. Some mountains in Saudi Arabia are almost 10,000 feet high. A large plateau covers the central part of the country. In the east, hills and plains cover the land near the Persian Gulf.

Deserts cover most of Saudi Arabia. The country has a very hot desert climate. The country has no lakes and no rivers. A large sand desert is in the south. It is called the Empty Quarter. Some of its **sand dunes** are 1,000 feet high.

The southwest has fertile soil. It is the only region that gets enough rain for farming.

Saudi Arabia covers most of the Arabian Peninsula. What three bodies of water surround the peninsula?

Saudi Arabia's flag

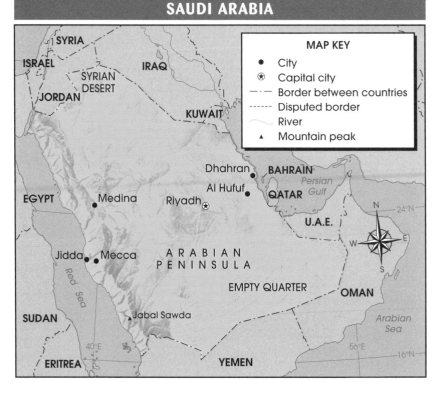

SAUDI ARABIA

SYRIA
ISRAEL
IRAQ
SYRIAN DESERT
JORDAN
KUWAIT
EGYPT
Dhahran
Al Hufuf
BAHRAIN
Persian Gulf
QATAR
Medina
Riyadh
U.A.E.
ARABIAN PENINSULA
Jidda
Mecca
Red Sea
EMPTY QUARTER
OMAN
SUDAN
Jabal Sawda
Arabian Sea
ERITREA
YEMEN

MAP KEY
● City
⊛ Capital city
–·– Border between countries
----- Disputed border
River
▲ Mountain peak

N 24°N
W E
S
40°E 56°E 16°N

Sand dunes in Saudi Arabia

Most people in Saudi Arabia live in cities. Riyadh, the capital, is near the center of the country. This city has almost 2 million people. It was built at an oasis in the desert. Jidda is a large port city on the Red Sea. Saudi Arabia also has two cities that are holy to Muslims. Mecca is the most important city to Muslims because Muhammad was born there. Medina is the other holy city.

History, People, and Government

Arabs have lived in Saudi Arabia for thousands of years. Many of the people were Bedouins. Bedouins are nomads. They move from one oasis in the desert to another. They travel on camels, live in tents, and raise sheep and goats. Many Bedouins still live in the Saudi deserts today. Some modern Bedouins travel by car instead of on camels.

About the year 570, Muhammad was born in Mecca. He started the religion of Islam. During his life, people living in the Arabian Peninsula became Muslims. Since that time Saudi Arabia has been a Muslim country. More than one billion people in the world today are Muslims.

Today about 19 million people live in Saudi Arabia. The people are called Saudis. Arabic is the official language. One fourth of the people are **foreign** workers, or workers from other countries.

King Fahd is the ruler of Saudi Arabia. He is a member of the royal Saudi family. This family has been important for hundreds of years. Saudi Arabia is not a democracy. It is an **absolute monarchy**. This means the king has full power to make all laws. The king is also the country's religious leader.

Religion and Women

All citizens of Saudi Arabia must be Muslims. They cannot practice any other religion. All Saudi laws are based on Islamic law. There are special police officers who make sure people obey the Islamic laws. Islam allows Saudi men to have four wives. Islam does not allow people to eat pork or drink alcohol. There are no movie theaters, plays, or concerts in the country.

There are strict Islamic laws for women. Girls must study at separate schools from boys. Women are not allowed to drive cars. They cannot ride bicycles. When women leave their homes, they must always be with a male family member. Women cannot work outside of their homes at jobs with men. They can only work with women. Some women work as teachers at schools for

King Fahd

A computer class in Riyadh

Every year Muslim men and women from around the world make a Hajj to the Great Mosque at Mecca.

A Saudi woman in an aba

girls. Other women work as doctors and nurses, but only with female patients.

Islamic laws say women must be **modest**. They must wear clothes that cover their arms and legs. They must cover their hair and faces. So Saudi women have to wear a long black robe called an **aba** when they leave home. An aba covers a woman's head, face, and body.

Muslims believe they must visit the holy city of Mecca at least once during their lives. The visit to Mecca is called a **Hajj**. Muslim men and women from all over the world make this religious trip to Mecca. There they visit the Great Mosque. This huge mosque holds 300,000 people. Inside the Great Mosque is the **Kaaba**, the holiest place in Islam.

Economy and Standard of Living

Oil and natural gas were discovered in Saudi Arabia in the 1930s. One fourth of all the oil in the world is in Saudi Arabia. Most of it is near the Persian Gulf. Saudi Arabia also has other minerals. It has iron, gold, and copper. However, Saudi Arabia has not developed these other natural resources.

Since the 1940s Saudi Arabia has earned billions of dollars from exporting oil. Saudi Arabia is a member of **OPEC**, the Organization of Petroleum Exporting Countries. Saudi Arabia is one of the richest countries in the world. But Saudi Arabia is still a developing country.

Many Saudis worry that some day they will have no more oil to export. Saudis want their economy to depend less on oil exports. So the country is finding other ways to earn money. It has a favorable balance of trade because it earns so much money from selling oil. It buys cars, food, machines, weapons, and many other products from the United States. But Saudi Arabia is also using its money from oil to build a modern country and to help its people.

About one fourth of the Saudis work at agriculture. The government has built **desalination plants** to take the salt out of ocean water. This water is used to grow food in the desert. Saudi Arabia now grows more than enough wheat. But it must still import more than half of its food.

Saudi Arabia also uses the money from oil to build an economy that is less dependent on selling oil. It is building modern factories and new industries. Today Saudis produce steel, cement, and food products. The country also earns millions of dollars from Muslims from other countries who come to Mecca for the Hajj.

The Saudis are working to raise the country's standard of living. Small villages now have electricity. Cities now have many modern apartment houses. There is free health care for all people. There are free public schools for children everywhere. Still, more than one third of the Saudi people cannot read and write.

Saudi Arabia is working to become a modern, industrial nation. It is already an important nation in the Middle East. It is also a powerful leader in OPEC and the Muslim world.

An oil refinery in Saudi Arabia

Desalination plants like this take the salt out of ocean water.

Chapter Main Ideas

1. Saudi Arabia is a desert country. It has one fourth of the world's oil.

2. All Muslims must visit Mecca on a Hajj. All Saudis must be Muslims.

3. Saudi Arabia is using money from selling oil to become a developed country. It is improving health care and education.

◆ Vocabulary

Finish the Paragraph   Use the word or words in dark print to finish the paragraph below. On your paper write the word or words you choose.

Hajj   **modest**   **absolute monarchy**   **desalination plants**
aba   **Kaaba**   **sand dunes**

The Bedouins of Saudi Arabia live in the desert and move from place to place. In the Empty Quarter Desert, there are large hills called __1__. The Saudis have built __2__ to change ocean water to fresh water. The government of Saudi Arabia is an __3__ because the king makes all laws. Saudis believe women must cover their hair, faces, arms, and legs in order to be __4__. City women wear a long black robe called an __5__. All Muslims try to make a religious trip to Mecca called a __6__. The holiest place in Mecca is the __7__.

◆ Read and Remember

Complete the Chart   Copy the chart shown below on your paper. Use facts from Chapters 33 and 34 to complete the chart. You can read both chapters again to find facts you do not remember.

Two Countries of the Middle East

	Saudi Arabia	Turkey
What is the official language?		
What is the religion?		
Is there religious freedom?		
What is the type of government?		
What are the resources?		
Is this a developing country?		
How does the country earn money?		

◆ Think and Apply

Drawing Conclusions   Read each pair of sentences. Then look in the box for the conclusion you might make. Write the letter of the conclusion on your paper.

1. Saudi Arabia gets almost no rain.
   The country has no lakes or rivers.

   Conclusion: _____

2. All Muslims must try to visit Mecca at least once.
   The holiest places in Islam are in Mecca.

   Conclusion: _____

3. It is against the law in Saudi Arabia to eat pork or drink alcohol.
   It is against the law for women to drive cars in Saudi Arabia.

   Conclusion: _____

4. Saudi Arabia uses money from selling oil to build schools and desalination plants.
   Saudi Arabia uses money from selling oil to bring electricity to small villages.

   Conclusion: _____

5. Saudi Arabia is building new factories.
   Saudi Arabia is earning money from its farm products.

   Conclusion: _____

Conclusions
   A. Saudi laws are based on Islamic law.
   B. Saudi Arabia is using money from oil to become a developed country.
   C. Saudi Arabia wants other ways to earn money besides selling oil.
   D. Mecca is a very holy city to Muslims.
   E. Most of Saudi Arabia is desert.

◆ Journal Writing

Write a paragraph in your journal that tells how Saudi Arabia is using the money it earns from selling its oil.

Reviewing Bar Graphs

You have learned that a **bar graph** uses bars of different lengths to show facts. This bar graph shows how much oil was produced by five countries in the Middle East in 1995. The amount of oil is measured in barrels.

Study the graph. Then write the answer to each question on your paper.

1. Which country produced the most oil?

   Kuwait   Algeria   Saudi Arabia

2. Which country is the second largest oil producer in the Middle East?

   Kuwait   Iran   Libya

3. Which countries produced less oil than Iran?

   Kuwait and Saudi Arabia

   Algeria and Libya

   Saudi Arabia and Algeria

4. About how many barrels of oil a day did Saudi Arabia produce in 1995?

   2,000   3,000   8,000

5. About how many barrels of oil a day did Libya produce in 1995?

   1,000   3,000   4,000

6. Which country produced the least oil?

   Algeria   Kuwait   Libya

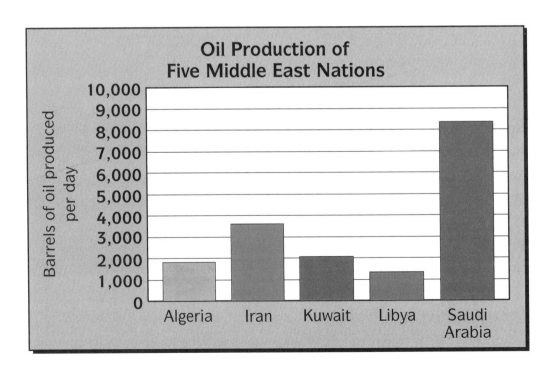

CHAPTER 35

Understanding the Middle East

Where Can You Find?
Where can you find a city that is so crowded that some people must live in a cemetery?

Think About As You Read

1. How are countries solving the problem of scarce water?
2. How has Islamic fundamentalism changed the Middle East?
3. How have wars hurt the Middle East?

New Words
- overpopulation
- cemetery
- aquifers
- conservation
- shah
- missiles

People and Places
- Kuwait
- Syria
- Iraq
- Iran
- Lebanon
- Saddam Hussein
- Queen Noor
- King Hussein

Most of the countries in the region of the Middle East are developing countries. To help their people have better lives, these countries must solve five problems.

Overpopulation, Poverty, and Lack of Water

The first problem in the Middle East and North Africa is **overpopulation**. This means there are too many people living in the region. Many cities are very crowded. Most of the land cannot be farmed. So most countries cannot grow enough food for all their people. Cairo, Egypt, has become so crowded that many poor people now live in a large **cemetery**.

The second problem is poverty. Some countries earn billions of dollars from exporting oil, but most people are poor. Too many people work as subsistence farmers. This region needs more factories and industries.

The third problem is the lack of water in the Middle East. It is the driest region in the world. Some countries, such as Saudi Arabia and Kuwait, have built desalination plants to solve this problem. Saudi Arabia

Many poor people live in this cemetery in Cairo.

A desalination plant

has more of these plants than any other nation in the world. But it is very expensive to change salt water to fresh water. Saudi Arabia and Kuwait use their money from selling oil to pay for the plants. Poor countries do not have enough money for desalination.

Other countries are building dams that save river water in reservoirs. Turkey and Syria have built dams on the Euphrates River. But these dams collect so much water that less river water flows into Iraq. Iraq is angry that it gets less water from the Euphrates.

In Libya water has been found in **aquifers** in the Sahara. An aquifer is an area of underground water. Pipes have been built to carry this water to cities near the coast.

In Israel people use **conservation** to save water. People must use as little water as possible for cooking, bathing, and cleaning. Farmers use drip irrigation to save water. Factory water is sometimes recycled. The government strictly controls the use of water.

Islamic Fundamentalism and War

The fourth problem is the growth of Islamic fundamentalism. One goal of this movement is to win control of governments in the region. The new governments would be based on strict Islamic law. In 1979 Islamic fundamentalists led a revolution in Iran.

The **shah**, or king, of Iran was forced to leave the country. The nation became an Islamic republic. Now everyone in Iran must obey strict religious laws.

Islamic fundamentalists have tried to win control of Egypt. Terrorists from the movement have bombed cities along the Nile River. In 1995 they tried, but failed, to kill Egypt's president. Many people fear that the fundamentalists will continue to spread terrorism.

War is the fifth problem in the Middle East. In Lebanon, fighting between Muslims and Christians led to a long civil war. Large areas of the country were destroyed. Most of the fighting ended in 1991.

Between 1980 and 1988, Iran and Iraq fought a long war. The fighting has ended, but the nations have not signed a peace treaty.

In 1990 soldiers from Iraq took control of Kuwait, Iraq's southern neighbor. Saddam Hussein, Iraq's president, wanted Kuwait's oil fields and its ports on the Persian Gulf. Hussein said Iraq would also attack Saudi Arabia. Iraq refused to leave Kuwait. So in 1991 the United Nations sent soldiers to force Iraq to leave Kuwait. The United States led soldiers from 28 countries in the war to free Kuwait. That war is called the Persian Gulf War. Iraq lost the war, and Kuwait became free again. But Hussein continues to rule Iraq and to make threats against Iraq's neighbors.

Since 1948 the Arab nations have been at war with Israel. Egypt and Jordan are the only Arab countries that have signed peace treaties with Israel. During the Persian Gulf War, Israel did not fight against Iraq. But Iraq fired **missiles**, or weapons with bombs, at Israel. The missiles damaged Israeli cities.

In 1993 the United States helped Israel and the PLO sign a peace agreement. Today the PLO controls part of the land the Palestinians lost in 1948. The two sides do not agree about how much land the PLO should control. But both sides are trying to find a peaceful way to live together in the region.

Most children in the Middle East have better lives than their grandparents had. There are more schools

Many buildings were destroyed during the civil war in Lebanon.

British soldiers in Kuwait in 1991

A new hotel being built in Iraq

and more factories. There is better health care. Some nations are using their money from selling oil to build roads, hospitals, and apartment houses. The people of this region need peace. Then they can work together to solve their problems.

Chapter Main Ideas

1. Overpopulation and poverty are problems in the Middle East. There is not enough farmland to grow food for all the people.
2. Water is scarce. Desalination plants and water conservation are helping the region.
3. There have been many wars in the Middle East. This region needs peace to solve its other problems.

BIOGRAPHY

Queen Noor of Jordan (Born 1951)

Lisa Halaby grew up as a Christian in a rich American family. She studied at Princeton University. In 1978 Halaby married King Hussein of Jordan. She became his fourth wife. She also became a Muslim. Halaby's name became Queen Noor.

The people of Jordan like Queen Noor because she works hard to help their country. She has helped Jordan have better schools. She has helped women and children have better lives. The queen also works with groups that help poor people.

The people of Jordan admire the way Queen Noor has cared for the three children of the king's third wife, Alia. Alia died in an accident. Queen Noor also has four children of her own.

Islamic fundamentalists want Queen Noor to wear traditional Arab clothes. But the queen wears western clothes while she works for Jordan and King Hussein.

Journal Writing
Write a paragraph in your journal about Queen Noor. Tell how her life changed after she married King Hussein.

USING WHAT YOU'VE LEARNED

◆ Vocabulary

Match Up   Finish the sentences in Group A with words from Group B. Write the letter of each correct answer on your paper.

Group A

1. Underground water is found in _____.

2. Programs for saving water are called _____.

3. The king of Iran was the _____.

4. Weapons with bombs are _____.

5. If there are too many people in a region, there is _____.

Group B

A. missiles

B. shah

C. aquifers

D. overpopulation

E. conservation

◆ Read and Remember

Finish Up   Choose the word or words in dark print that best complete each sentence. Write the word or words on your paper.

| Nile River | aquifers | Persian Gulf War | Iraq |
| Israel | Iran | United States | |

1. Recycling factory water and using drip irrigation are two examples of water conservation in _____.

2. Libya gets water from underground _____.

3. Islamic fundamentalists have bombed cities near the _____.

4. Iraq and _____ fought an eight-year war over several conflicts.

5. The _____ started after Iraq attacked Kuwait.

6. _____ lost the Persian Gulf War.

7. The _____ helped Israel and the PLO reach a peace agreement.

Cause and Effect Number your paper from 1 to 7. Write sentences on your paper by matching each cause on the left with an effect on the right.

Cause

1. Most of the Middle East cannot be farmed, so ———.

2. Water is scarce in Saudi Arabia and Kuwait, so ———.

3. Turkey and Syria have dams on the Euphrates River, so ———.

4. In 1979 Iran became an Islamic republic, so ———.

5. Egypt has made peace with Israel, so ———.

6. Iraqi soldiers would not leave Kuwait, so ———.

7. Iraq fired missiles at Israel during the Persian Gulf War, so ———.

Effect

A. they use desalination plants to make fresh water

B. the United States led soldiers in the Persian Gulf War

C. the shah was forced to leave his country

D. Israeli cities were damaged

E. Iraq is getting less water

F. countries do not grow enough food for their people

G. Islamic fundamentalists commit terrorist acts against both countries

◆ **Journal Writing**

You have been reading about the Middle East. Most of the countries of this region are developing countries. They have problems they need to solve to help their people have better lives. What are two problems in the Middle East that you think are most important? Write a paragraph that explains the two problems. Then tell how they might be solved.

South and Southeast Asia

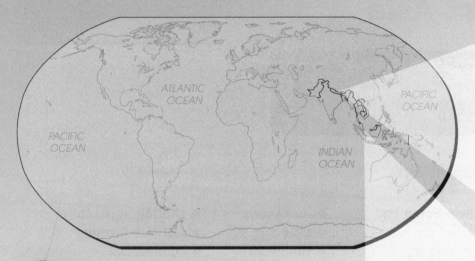

Mount Everest

DID YOU KNOW?

▲ Vietnam is a country with more than 200 rivers.

▲ Singapore is both a city and a nation. It is about the size of Chicago.

▲ Rice has been grown in this region for more than 5,000 years.

▲ The game of chess was first played in India.

▲ The king of Thailand has ruled his country since 1945. No other country in the world today has a monarch who has ruled for so many years.

▲ The tiny country of Brunei is so rich from exporting oil that the people of Brunei do not pay any taxes.

Singapore

WRITE A TRAVELOGUE

Look at the photographs in Unit 7. Then choose two of the countries you would like to visit. In your travelogue, write about what you would like to do in those countries. After reading Unit 7, write two or more paragraphs that describe how South Asia and Southeast Asia are regions.

THEME: REGION

287

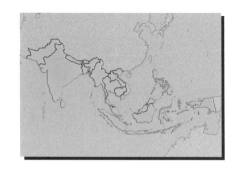

CHAPTER 36

Getting To Know South and Southeast Asia

Where Can You Find?
Where can you find the tallest mountain in the world?

Think About As You Read

1. How do the monsoons help South and Southeast Asia?
2. What kind of land and climate do South and Southeast Asia have?
3. How do most people earn a living?

New Words

- ◆ monsoons
- ◆ timber
- ◆ Hinduism
- ◆ Buddhism
- ◆ subcontinent
- ◆ shifting agriculture
- ◆ wet rice farming
- ◆ commercial agriculture

People and Places

- ◆ Himalaya Mountains
- ◆ Mount Everest
- ◆ India
- ◆ Indonesia
- ◆ Brunei
- ◆ the Netherlands
- ◆ Thailand
- ◆ Pakistan
- ◆ Bangladesh
- ◆ Hindus
- ◆ Philippines
- ◆ Deccan Plateau
- ◆ Indo-Gangetic Plain
- ◆ Indus River
- ◆ Ganges River
- ◆ Vietnam
- ◆ Mekong River

If you lived in South or Southeast Asia, you might live in a small village. You would probably work on a farm. Rice would be your most important food.

Landforms, Climates, and Resources

The land of South and Southeast Asia is located between the Indian and Pacific oceans. This region has two parts. One part is South Asia. The other part is Southeast Asia. Most of the region is covered with plains and plateaus. The world's tallest mountain chain separates South Asia from the rest of Asia. These tall mountains are the Himalayas. One of these mountains, Mount Everest, is the tallest mountain in the world.

Most of this region is in the tropics. It has a hot climate. **Monsoons**, or seasonal winds, are part of the region's climate. Most of the countries in this region get monsoon winds. From April until October monsoon winds blow in one direction. They bring heavy rain during the summer. Farmers need this rain to grow

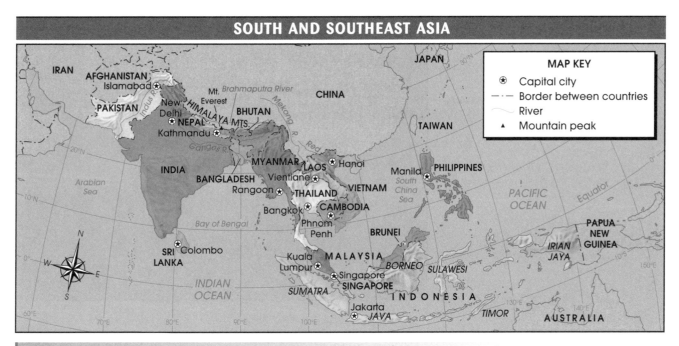

The region of South and Southeast Asia has many islands. The western part of one island belongs to Indonesia. What is the western part of that island called?

food. The winds blow in a different direction for the rest of the year. The winter monsoons bring dry air. There is little or no rain in winter.

South and Southeast Asia have important natural resources. India has coal. Indonesia and Brunei have oil. Some countries get rubber and **timber** from the trees in the region's tropical rain forests. Timber is wood that is used to make buildings and furniture. Countries in the region export some of their resources. But they do not use their resources to make factory goods. Most of the countries in this region are developing nations.

History and People

In the 1500s people from Europe won control of most of the region. For hundreds of years, Britain, France, and the Netherlands ruled many colonies here. Only Thailand, in Southeast Asia, remained free. Now all of the countries in this region are free.

The European countries sent resources from their Asian colonies back to Europe. They used these resources to make factory goods. Europeans also started large plantations to grow cash crops. Today the countries of the region continue to grow such cash

The Himalaya Mountains

A rubber plantation

crops as rubber, tea, cotton, sugar, and spices. They export these crops.

There are many ethnic groups in South and Southeast Asia. The people of this region speak many different languages. They practice different religions. Most people in Pakistan, Bangladesh, and Indonesia are Muslims. Most Indians are Hindus. The religion of **Hinduism** began in India thousands of years ago. The religion of **Buddhism** also began in India long ago. From India it moved to Southeast Asia. People practice Buddhism in many Southeast Asian countries. In the Philippines most people are Christians. There have been many conflicts between people of different religions in this region.

Looking at South Asia

More than one billion people live in South Asia. The area is more crowded than Southeast Asia.

South Asia is on a large peninsula. It is called a **subcontinent** because it is a very large piece of land, but it is smaller than a continent. The subcontinent has three parts. One part is the Himalaya Mountains in the north. The second part is the dry Deccan Plateau. There are deserts on the plateau. The Indo-Gangetic Plain is the third part of South Asia. It is south of the Himalayas. This plain surrounds the Indus and Ganges rivers. These rivers start high in the Himalaya Mountains. The land around the rivers is very fertile.

Many of the cities in South Asia are crowded.

It is densely populated. For thousands of years people have lived in these fertile river valleys.

India is the largest country in South Asia. Pakistan is to the west. Bangladesh is to the east. About three fourths of the people in South Asia live in villages. People live in small huts made of mud. Many villages do not have electricity, running water, cars, or telephones. There are also large cities with millions of people. Most of the people in these cities are very poor. The cities do not have enough jobs for all the people who need work.

The Mekong River in Vietnam

Most people in South Asia are subsistence farmers. They work on small farms using traditional methods. As soon as the summer monsoons bring rain, people start planting crops. Rice is the most important crop. India grows almost enough food for its huge population. Pakistan and Bangladesh import large amounts of food.

Looking at Southeast Asia

Southeast Asia is a crossroads region. For thousands of years people have passed through Southeast Asia as they traveled between Europe, the Middle East, Africa, and Asia. Indonesia, Vietnam, and the Philippines are three of the many countries in Southeast Asia.

Southeast Asia has many islands and peninsulas. It has a long seacoast. Most people live near the sea. There are important rivers in the area. One river, the Mekong, is the fifth longest river in the world.

Southeast Asia has a tropical climate. The monsoons bring summer rains. The hot, wet climate allows tropical rain forests to grow.

Chopping down trees destroys the rain forests.

Southeast Asia has some very large cities. But most people live in villages. They work at farming.

There are three types of farming in Southeast Asia. The first type is **shifting agriculture**. People chop down trees in the forests and burn them. Then they plant crops. After several years, the soil is no longer fertile. So people move to another part of the forest. There they chop down trees to farm new land. This method destroys the rain forests. It also causes deforestation.

Wet rice farming in Indonesia

Picking tea leaves

The second type of farming is **wet rice farming**. Rice seeds are planted in small flooded fields after heavy rains. This work is done by hand. Many farmers are needed. But much of the rice in South and Southeast Asia is grown this way. Rice is the region's main food.

Commercial agriculture is the third type of farming. People grow cash crops on large plantations. Countries export cash crops such as tea and cotton to earn money.

The countries of South and Southeast Asia are working to become more developed. As you read this unit, find out how countries in the region are changing.

Chapter Main Ideas

1. Summer monsoons bring rain to the region. Farmers need this rain to grow food.
2. South and Southeast Asia have huge populations. Most people are subsistence farmers.
3. People have lived in this region for thousands of years. All of the countries except Thailand were once European colonies.

◆ Vocabulary

Forming Word Groups   Copy the chart shown below on your paper. Form groups by writing each vocabulary word under the correct heading on your paper. There is one word you will not use.

Vocabulary List

wet rice farming	commercial agriculture	Buddhism
Hinduism	monsoons	shifting agriculture
subcontinent	timber	

Understanding South and Southeast Asia

Religions	Land and Climate	Types of Farming
1. _____	1. _____	1. _____
2. _____	2. _____	2. _____
		3. _____

◆ Read and Remember

Complete the Geography Organizer   Copy the geography organizer and complete it with information about South and Southeast Asia.

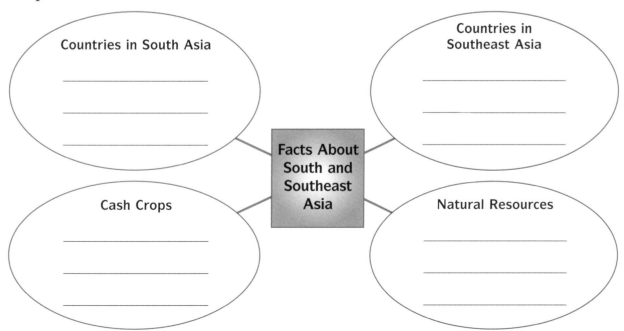

Write the Answer   Write one or more sentences to answer each question.

1. How do the monsoons help South and Southeast Asia?

2. What did European countries do in this region?

3. Which country in Southeast Asia has never been a colony?

4. What cash crops do the countries of this region export?

5. How do most people earn a living in South and Southeast Asia?

6. What important natural resources can be found in this region?

◆ Think and Apply

Fact or Opinion   Number your paper from 1 to 10. Write **F** next to each fact. Write **O** next to each opinion. You should find four sentences that are opinions.

1. The Indus, Ganges, and Mekong rivers are important to South and Southeast Asia.

2. The Himalaya Mountains are the tallest in the world.

3. Monsoon winds bring dry air in winter.

4. The region needs more plantations.

5. European nations began controlling South and Southeast Asia in the 1500s.

6. Most people in South and Southeast Asia are poor farmers.

7. South and Southeast Asia should export more resources.

8. There should be laws to stop people from chopping down rain forests.

9. Many people in South Asia live on the fertile plain around the Ganges and Indus rivers.

10. More people should move to the large cities of South and Southeast Asia.

◆ Journal Writing

Write a paragraph in your journal that explains why the summer monsoons are important. Tell how people use the rain.

CHAPTER 37

India: Largest Nation of South and Southeast Asia

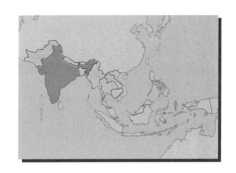

Where Can You Find?
Where can you find a crowded eastern port city with a modern subway?

Think About As You Read

1. How do problems with the monsoons hurt India?
2. How does the Hindu religion affect Indian life?
3. How is India becoming a more developed nation?

New Words

- ◆ sacred
- ◆ Hindi
- ◆ Sikhism
- ◆ vegetarians
- ◆ caste system
- ◆ caste
- ◆ untouchables
- ◆ extended families
- ◆ jute
- ◆ religious order
- ◆ orphanages
- ◆ orphans

People and Places

- ◆ Ganges Plain
- ◆ Brahmaputra River
- ◆ Sikhs
- ◆ Mohandas Gandhi
- ◆ Bombay
- ◆ Calcutta
- ◆ New Delhi
- ◆ Delhi
- ◆ Taj Mahal
- ◆ Mother Teresa

India is the largest nation in South and Southeast Asia. It has the second largest population in the world. Almost one billion people live in India. It is a developing nation with big cities and thousands of villages.

India's Landforms, Climate, and Resources

The Himalaya Mountains separate India from China. The Ganges Plain is south of the tall mountains. It is part of the Indo-Gangetic Plain. The Ganges Plain covers northern India. Half of India's people live on this fertile plain. The dry Deccan Plateau is south of the Ganges Plain.

Three rivers flow through the Ganges Plain. They are the Indus River, the Ganges River, and the Brahmaputra River. The Ganges is the longest river. It is **sacred**, or holy, to Hindus. Many Hindus bathe in the holy water of the Ganges River.

Southern India is a large peninsula. It is mostly covered by the Deccan Plateau. There are narrow

Mount Everest is the tallest mountain in the world. Where is it located?

India's flag

INDIA

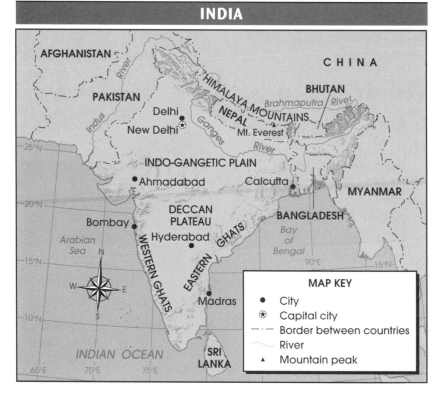

MAP KEY
- City ●
- Capital city ⊛
- Border between countries – · –
- River ～
- Mountain peak ▲

It is often difficult to cross a river after a monsoon.

plains along the coasts. There are low mountain chains in the east and west near the coastal plains.

India has three climate seasons. One season is cool and dry. Another is hot and dry. The monsoons make June to September a hot and rainy season.

Sometimes the monsoons do not bring enough rain. Then there are droughts. Farmers cannot grow enough food. Sometimes the monsoons bring too much rain. Then the country has dangerous floods.

India has many natural resources. It has coal, iron ore, bauxite, and other minerals. Waterpower from rivers is used to make electricity. India also has some oil. But India imports a lot of the oil that it needs.

People, Religion, and Culture

The people of India are called Indians. They speak 16 main languages. It is hard for people to understand each other because there are so many languages. **Hindi** is the main official language, but only about one third of the people speak it. English is used for government work and business.

Religion is very important to the people of India. Most people believe in the religion of Hinduism. These people are called Hindus. The country also has millions of Muslims. There are also Christians. Other people follow an Indian religion called **Sikhism**. People who follow this religion are Sikhs. All people have religious freedom in India.

Hinduism is a way of life in India. Hindus pray to different gods. They believe it is wrong to hurt people or animals. Cows are sacred to Hindus. There are many cows on city streets. Hindus drink milk from cows, but they do not eat meat from cows. Many Hindus are **vegetarians**, people who do not eat any meat.

The **caste system** is part of Hinduism. A **caste** is a group of people. Each Hindu is born into a caste. People cannot change their caste. Hindu priests and their families are in the highest caste. Farmers are in a lower caste. People of the same caste live near each other in a village. People must marry a person from their own caste.

A Hindu priest

One group of people does not belong to any of the castes. They are called **untouchables**, people who cannot be touched. They have always done the dirtiest jobs. In 1950 the Indian government gave equal rights to the untouchables and to all lower castes.

Religion has caused fighting in India. There have been many fights between Hindus and Muslims. There have also been fights between Hindus and Sikhs.

Family life is an important part of India's culture. Many people live together in **extended families**. This

Cows can often be found on city streets in India.

A boat on the Indus River

Mohandas Gandhi

means grandparents, aunts, uncles, children, and parents live together. Parents often decide who their children will marry.

History and Government

About 4,500 years ago, people first settled on the fertile land near the Indus River. People later moved into southern India.

In the 1800s Great Britain forced India to become a British colony. The British built railroads, roads, and schools. But the Indians wanted to rule themselves.

After World War I, a great leader named Mohandas Gandhi led India's struggle for independence. Gandhi was against all fighting and violence. He used peaceful protests to force the British to leave India. At last, in 1947 India became an independent country. That same year the eastern and western parts of India became a Muslim nation called Pakistan. Millions of Indian Muslims in India moved to Pakistan. Millions of Hindus in Pakistan moved to India. Today eastern Pakistan is an independent country called Bangladesh.

India is the world's largest democracy. People vote for Parliament members. A prime minister leads the country.

Cities, Villages, and the Economy

Only one fourth of India's people live in cities. About 13 million people live in Bombay. This busy port is the largest city in India.

Calcutta is a large port in the east. Calcutta has a modern subway system. It has many kinds of factories. But about one fourth of the people live in slums. Thousands of homeless, starving people live in the streets.

India's capital is in the modern city of New Delhi. An older city called Delhi is next to New Delhi.

One of the most beautiful places in the world is not too far from Delhi. It is the famous Taj Mahal. An Indian emperor had it built in the 1600s to honor his dead wife.

The Taj Mahal

Most Indians live in small villages. There are more than 500,000 villages. Most villages have about 1,000 people. People live in small huts. Most villages do not have electricity or running water. So women must go to the village well to get water for their families. There they learn the latest news from other women. Women carry water in pots on their heads back to their homes. Most village people are poor subsistence farmers. They do not have modern machines or tools. Their main foods are wheat or rice and vegetables.

Indian farmers also grow cash crops on large plantations. India exports cotton, sugar, tea, spices, and **jute**. Jute is used to make rope.

India is becoming an industrial nation. Millions of city people have factory jobs. Today India produces cars, ships, planes, steel, machines, and many other products.

India has a large movie industry. Indians make more than 700 movies each year. The country also earns money from tourism.

A woman carrying water

A Changing Nation

India is working hard to become a developed nation. Illiteracy has been a big problem. About half of the people cannot read. So the government has started many new schools. Most small villages now have an elementary school.

For many years India did not grow enough food. The government has helped farmers learn better ways to grow food. You will read more about this in Chapter 40. Today India grows almost all the food it needs.

India's biggest problem is rapid population growth. To solve this problem, the government gives extra money to families that have only two children.

Indians are building more schools and factories. They are learning better ways to farm. They are solving the problems of their country.

Chapter Main Ideas

1. Hinduism and the caste system are a way of life in India.
2. In 1947 India became free from Great Britain. The government is a democracy.
3. Most Indians live in villages. One fourth of the people live in large cities.

BIOGRAPHY

Mother Teresa (1910-1997)

Mother Teresa grew up in Eastern Europe. There she became a Catholic nun. She was sent to Calcutta, India. Mother Teresa started a **religious order** called the Missionaries of Charity. Other nuns joined the order. They helped Mother Teresa care for starving, dying, sick, and homeless people in Calcutta. Their job was difficult because so many people needed their help.

Mother Teresa started schools and hospitals. She opened **orphanages** to care for **orphans**. She started branches of her organization in many Indian cities. She also started branches in other countries.

In 1979 Mother Teresa won the Nobel Peace Prize. She continued helping India's poorest people. Her health was poor the last years of her life. She gave her job of running the organization to another nun. Mother Teresa is loved by the Indian people because she helped their country for so many years.

Journal Writing
Write a paragraph in your journal that tells how Mother Teresa has helped India.

p 324 - Singapore

1. It is an island nation with a tropical climate. It is a rich, developed nation with a busy trade port.

2. Most of the animals living on Singapore were destroyed when roads and factories were built.

3. Singapore is a region of very strict government.

4. Sinapore is in Southeast Asia near Malaysia and Indonesia. It is very close to the Equator.

5. The port of Singapore is used by ships from many nations as they travel the trade routes

◆ Vocabulary

Finish Up Choose the word in dark print that best completes each sentence. Write the word on your paper.

Sikhism	**untouchables**	**orphanage**	**vegetarians**
Jute	**caste**	**Hindi**	**sacred**

1. The main official language of India is _____.

2. People who do not eat meat are _____.

3. A Hindu is born into a group called a _____.

4. _____ is a plant that is used for making rope.

5. One of the religions in India is _____.

6. People who do not belong to a caste are _____.

7. A group home for children without parents is an _____.

8. Something that is holy is _____.

◆ Read and Remember

Write the Answer Write one or more sentences to answer each question.

1. Where is the Ganges Plain?

2. How do the monsoons help and hurt India?

3. What are India's natural resources?

4. What four main religions do people in India follow?

5. How do people get water in many villages?

6. Who led India's struggle for independence after World War I?

7. Why was the Taj Mahal built?

8. What are the two largest port cities in India?

9. What are some of India's factory products?

10. How does the caste system affect life in India?

11. What kind of government does India have?

12. What is India's climate like?

13. What are the two main languages spoken in India?

14. Where do most people in India live?

◆ Think and Apply

Finding Relevant Information Imagine you are telling your friend why India is a developing country. Read each sentence below. Decide which sentences are relevant to what you will say. On your paper write the relevant sentences you find. You should find six relevant sentences.

1. Most people live in small villages.

2. Illiteracy is a big problem.

3. India exports cotton, spices, tea, and jute.

4. Indian factories produce cars, steel, and other products.

5. Most people are subsistence farmers.

6. Many villages do not have electricity or running water.

7. One fourth of Calcutta's people live in slums.

8. India grows almost enough food for its population.

9. Monsoons bring rain from June to September.

10. Farmers use work animals instead of modern farm machines.

◆ Journal Writing

Only one fourth of India's people live in cities. Most Indians live in small villages. How is life in an Indian village different from the place where you live? Write a paragraph in your journal that tells how the places are different from each other. Give at least three examples.

Reviewing Circle Graphs

A **circle graph** is a circle divided into parts. All the parts form a whole circle. Look at the circle graph on this page. It shows the percent of Indians in each religious group. The graph shows how the total population is divided among five groups of people.

Use the words in dark print to finish the sentences. Write the words on your paper.

Christians   **Muslims**   **Hindus**   **1%**   **Sikhs**   **11%**   **3%**

1. The _____ are 2% of the population.

2. The Muslims are _____ of the population.

3. About _____ of the population is Christian.

4. The graph shows that the _____ are the largest religious group.

5. The _____ are the second largest religious group.

6. The percent of Sikhs plus the group labeled Others is equal to the percent of _____.

7. Buddhism began in India. From the graph we can conclude that Buddhists are less than _____ of the population of India.

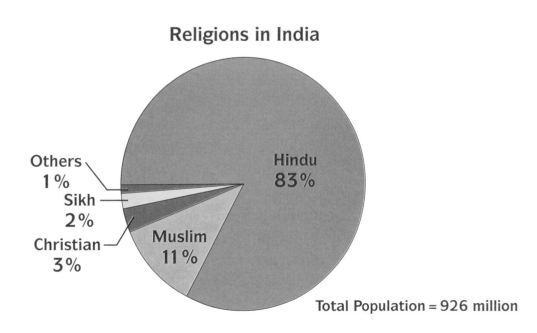

Religions in India

Others 1%
Sikh 2%
Christian 3%
Muslim 11%
Hindu 83%

Total Population = 926 million

CHAPTER 38

Vietnam: A Changing Country in Southeast Asia

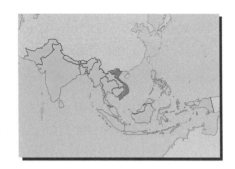

Where Can You Find?

Where can you find a large river delta in southern Vietnam?

Think About As You Read

1. Where do most people in Vietnam live?
2. How has Vietnam changed since 1975?
3. How do people earn a living in Vietnam?

New Words

- typhoons
- Tet Festival
- parallel
- Viet Cong
- boat people
- Communist party
- average
- manufacturing
- textiles
- investing

People and Places

- Indochina Peninsula
- Laos
- Cambodia
- Red River
- Red River delta
- Mekong River delta
- Ho Chi Minh City
- Hanoi
- Ho Chi Minh
- Geneva, Switzerland

Vietnam has always been a changing nation. As you read this chapter, think about the ways this Southeast Asian nation has changed during its long history.

Vietnam's Landforms, Climate, and Resources

Vietnam is on the eastern part of the Indochina Peninsula. China is north of Vietnam. Laos and Cambodia are west of Vietnam.

Vietnam's most important rivers are the Red River and the Mekong River. These rivers form two large deltas near the sea. The Red River delta is in the north. The Mekong River delta is in the south. There are coastal plains between the two deltas.

Most of Vietnam is covered with mountains. Forests and jungles cover the mountains.

Vietnam has a tropical climate. Monsoons bring very heavy rains between May and October. These heavy rains sometimes cause dangerous floods on the rivers. The climate is cooler in the mountains.

VIETNAM

Vietnam is located on the Indochina Peninsula. What three countries share borders with Vietnam?

Vietnam's flag

The central part of the coast often has **typhoons**. Typhoons are dangerous tropical Asian storms.

Vietnam has important resources. It has coal, iron, and bauxite. There is oil under the sea near the coasts. The country has not yet developed its resources.

People, Cities, and Culture

About 74 million people live in Vietnam. Most people are Buddhists. Some are Catholics. Vietnamese is the official language.

Most people live on the fertile deltas formed by the Red and Mekong rivers. Only one fifth of the people live in cities.

The largest city is Ho Chi Minh City in the south. It is close to the Mekong River delta. The city has more than 4 million people. It has the most factories and businesses. Hanoi is the nation's capital. It is a port on the Red River.

Families in Vietnam are very close. Members of extended families often live together. Parents and children live with grandparents, aunts and uncles, and cousins.

This family is making rice cakes to celebrate the Tet Festival.

American soldiers in Vietnam during the Vietnam War

The Vietnamese people celebrate the **Tet Festival**. It is the most important holiday. The Tet Festival celebrates the Vietnamese New Year. It is usually at the end of January. It is a happy time with parades and family visits.

History and Government

People first settled Vietnam about 3,500 years ago. Later, China ruled Vietnam for about 1,000 years. Vietnam became free in the year 939. It remained a free country until it became a French colony in the 1800s. The French started plantations that grew tea, rice, and rubber. They also built cities and railroads.

Ho Chi Minh

After World War II, Ho Chi Minh, a Communist leader in the north, wanted Vietnam to be free and independent. He led Vietnamese Communists in a war against France. The French lost the war in 1954. Peace talks were held in Geneva, Switzerland. During these talks, Vietnam was divided at the 17th **parallel**, or line of latitude. North Vietnam became a Communist country. South Vietnam became non-Communist.

Communists in North Vietnam soon began fighting to win control of South Vietnam. They were helped by Communists who lived in South Vietnam. Communists in South Vietnam were called the **Viet Cong**. Non-Communists in the south fought against the Viet Cong. This fight became the Vietnam War.

The United States sent American soldiers to help the South Vietnamese. By 1969 there were more than 500,000 American soldiers in South Vietnam. The war lasted many years. Thousands of people died. In 1973 North and South Vietnam agreed to stop fighting. Both sides signed a cease-fire agreement. The American soldiers returned home.

In 1975 Communists began fighting again for control of all of Vietnam. By April 30, 1975, they controlled the entire country. The north and the south became one Communist country in 1976.

Many non-Communists left Vietnam when the war ended. They often escaped in small boats like this.

At the end of the war, many non-Communists escaped from Vietnam in small boats. They were called **boat people**. These people became refugees in the United States, Australia, and other countries.

Vietnam's **Communist party** now controls the government. All government leaders are from this party. People are not allowed to speak or write against the government.

Standard of Living and Earning a Living

Vietnam has a very low standard of living. Most people are very poor. The **average**, or usual, salary is less than $100 a month. Few people own cars. People often travel on bicycles. People with more money buy motorcycles. Education is important in this developing country. Most people know how to read.

Most Vietnamese people work as subsistence farmers. Rice is a very important crop. Farmers grow rice on the flooded fields of the deltas. Vietnam exports a lot of rice. Since the country has a long coast, fishing is an important industry. There is some **manufacturing**. Some factory products are bicycles, cement, farm tools, and **textiles**. Textiles are different types of cloth.

These girls go to school on bicycles.

During the Vietnam War, tourists did not visit Vietnam. But now tourists are visiting the country again. Vietnam is earning millions of dollars from tourism. Many parts of the country were destroyed during the war. But people are slowly rebuilding their country.

Farmers selling their crops at an outdoor market

Foreign investments are helping Vietnam become more developed.

After the Vietnam War, the Communist government owned all of the businesses. Farms and factories earned little money. The country is changing to a free market economy. People can now own businesses.

For many years, there was no trade between the United States and Vietnam. But now the two nations trade with each other.

Vietnam does not have enough money to develop many new industries and factories. Business owners from Japan, the United States, and other developed countries are **investing** their money to develop Vietnam. They are using their money to start businesses. Business owners hope to earn a lot of money as Vietnam becomes more developed.

Vietnam has changed in many ways during its long history. It is now trying to become a nation with a higher standard of living.

Chapter Main Ideas

1. Most Vietnamese live on the Red River delta in the north and on the Mekong River delta in the south.
2. Vietnam was once ruled by China and later by France. Americans fought in the long Vietnam War.
3. The Communist party rules Vietnam. The country is changing to a free market economy.

◆ Vocabulary

Finish the Paragraph   Number your paper from 1 to 10. Use the words in dark print to finish the paragraphs below. Write the words you choose on your paper.

Communist party	**boat people**	**Tet Festival**
parallel	**Viet Cong**	

During the Vietnam War, South Vietnam's government was controlled by people who were against communism, or non-Communists. There were many Communists in South Vietnam who were called the __1__. Communists won control of all of Vietnam in 1975. The Vietnamese __2__, a political party controlled by Communists, won control of the country. Vietnam was no longer divided at the 17th __3__. Many people escaped from Vietnam in small boats. These people were called __4__. The Vietnamese usually celebrate the new year at the end of January. This important holiday is called the __5__.

average   textiles   manufacturing   investing   typhoons

Today Vietnamese factories are __6__ some products such as bicycles, cement, and farm tools. They also manufacture cloth, or __7__. Business owners from other countries are __8__, or using their money, to start new businesses in Vietnam. Vietnam is a poor country where the usual, or __9__, salary is less than $100 a month. Vietman has a tropical climate. Sometimes the central part of Vietnam's coast has __10__, or dangerous tropical storms.

◆ Read and Remember

Matching   Each item in Group A tells about an item in Group B. Write the letter of each item in Group B next to the correct number on your paper.

Group A

1. These form two deltas in Vietnam.

2. This Communist leader led the fight for independence from France.

3. After Vietnam defeated France, peace talks were held in this city.

4. This country had more than 500,000 soldiers in Vietnam in 1969.

5. This northern city is Vietnam's capital.

6. This is the largest city in Vietnam.

7. These are two of Vietnam's export crops.

Group B

A. Geneva

B. Hanoi

C. the Red River and the Mekong River

D. rice and rubber

E. Ho Chi Minh

F. Ho Chi Minh City

G. United States

◆ Think and Apply

Sequencing   Number your paper from 1 to 5. Write the sentences on your paper to show the correct order.

Vietnam became a French colony.

China ruled Vietnam for almost 1,000 years.

Vietnam was divided into two countries at the 17th parallel.

American soldiers fought with South Vietnam during the Vietnam War.

Vietnam became a united Communist country.

◆ Journal Writing

Vietnam is a nation that has changed many times. Write a paragraph in your journal that tells three ways that Vietnam has changed during its long history.

Reading a Statistics Table

We use many kinds of statistics, or information given in numbers, to describe a nation's standard of living. The literacy rate tells what part of the population can read. Life expectancy tells the average number of years people live. Average income tells about how much money people earn a year. Most people own telephones when there is a high standard of living.

Population and Standard of Living in Four Countries of South and Southeast Asia

	Average Income	Literacy Rate	Population	Telephones	Men's Life Expectancy
Bangladesh	$1,040	38%	123 million	one for 435 people	56 years
India	$1,360	52%	952 million	one for 112 people	59 years
Vietnam	$1,400	94%	74 million	one for 270 people	65 years
Singapore	$20,000	91%	3 million	one for 2 people	75 years

Study the statistics table. Then use the words and numbers in dark print to finish each sentence. Write the answers on your paper.

Bangladesh   **Singapore**   **life expectancy**   **$20,000**
low   **$1,500**   **Vietnam**   **literacy rate**

1. Vietnam's _____ of 94% is the highest on the table.

2. The lowest literacy rate and average income belong to _____.

3. The average yearly income in Singapore is _____.

4. After Singapore, the highest life expectancy for men is in _____.

5. The average yearly income in three countries is less than _____.

6. The highest standard of living is found in _____.

7. Countries with the highest literacy rate have a higher _____.

8. Developing nations have large populations and _____ standards of living.

CHAPTER 39

Indonesia: One Nation on Thousands of Islands

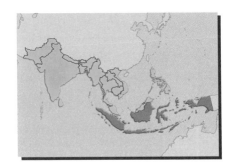

Where Can You Find?
Where can you find the second largest island in the world?

Think About As You Read

1. Why has it been hard for Indonesia to be a united country?
2. What are Indonesia's most important resources?
3. How has Indonesia's standard of living improved since 1968?

New Words

- archipelago
- ferry boats
- volcanic ash
- humid
- reelected
- Bahasa Indonesia
- stilts
- communicate

People and Places

- New Guinea
- Irian Jaya
- Papua New Guinea
- Sumatra
- Java
- Jakarta
- Sukarno
- Suharto

If you wanted to visit all of Indonesia, you would have to visit more than 13,670 islands. These islands cover 3,200 miles in the Indian and Pacific oceans. It has been difficult for people who live on so many islands to form a united country.

The Islands of Indonesia

Indonesia is an **archipelago**. An archipelago is a chain of islands. People live on about 6,000 islands in the chain. Six large islands have most of the nation's population. Some of the islands are so small they are not shown on maps. Many islands have no people on them.

The island of New Guinea is in eastern Indonesia. It is the second largest island in the world. Only the western part of New Guinea, Irian Jaya, belongs to Indonesia. The eastern part of the island is a different nation. It is called Papua New Guinea.

INDONESIA

Indonesia is an archipelago of many large and small islands. On which island can you find the capital city of Jakarta?

Sumatra is a large western island. It has important oil resources.

Java is the most important island in Indonesia. It has less than ten percent of the nation's land. But two thirds of the nation's people live on Java. Its soil is very fertile.

Jakarta is the country's capital and largest city. It is in northern Java. This city has about 9 million people. It is a busy port and trade center. It has many factories, banks, schools, and businesses.

It can be difficult to travel around Indonesia. **Ferry boats** help people travel between islands. Airplanes make it faster to travel from island to island. But most people do not have enough money for plane travel.

Landforms, Climate, and Resources

Indonesia has fertile plains near its coasts. Some of these plains have good harbors. There are also beautiful beaches. Mountains, hills, and plateaus cover parts of many islands.

The country has about 60 mountains that are active volcanoes. They can erupt at any time. When they erupt, lava and **volcanic ash** pour out. When this happens,

Indonesia's flag

A volcano on Java

On the streets of Jakarta

many people can be killed. Lava and volcanic ash also make the soil fertile for farming.

All of the islands of Indonesia are along the Equator. So the climate is always hot and **humid**.

Indonesia has the world's second largest tropical rain forest. The largest is in Brazil. Indonesians have chopped down large areas of their rain forest. They chop down trees to create new farmland. Trees are also cut down to be made into timber. The timber is exported to countries around the world. Many countries have told Indonesia to stop cutting down the trees. They want Indonesia to save its rain forests.

Indonesia is rich in resources. Oil is the most important resource. Indonesia also produces large amounts of tin and natural gas. The country also has bauxite, copper, gold, and coal.

Large parts of Indonesia's rain forests are being chopped down to create farmland.

History of Modern Indonesia

In 1816 Indonesia became a Dutch colony. The Dutch started plantations in the colony. They raised coffee, rubber, and other export crops. The Dutch did not share the money they earned with the island people.

After World War II, many Indonesians wanted their country to be independent. A leader named Sukarno led the fight for freedom. In 1949 Indonesia became a free nation. Sukarno ruled his country as a dictator. While he ruled, the country grew poorer.

Suharto, another leader, became president in 1968. He has been **reelected** five times. He continues to lead the country today.

People and Culture

About 207 million people live in Indonesia. The country has the fourth largest population in the world. The people belong to many ethnic groups. They speak more than 250 different languages. Indonesia has one official language called **Bahasa Indonesia**.

Indonesia is the world's largest Muslim country. Almost 90 percent of Indonesia's people are Muslims. There are many Christians too. There are also some Hindus and Buddhists.

About two thirds of the people live in villages. Most of the houses are built on long poles called **stilts**. The family's farm animals live in the space under the house.

It has been hard for the people of Indonesia to form a united nation. They speak many different languages. They live far apart on thousands of islands. It is hard for people on different islands to **communicate**. It is hard for people on different islands to visit each other. Sometimes mountains and forests separate people who live on the same island. The country is working to be united. All children study Indonesia's official language in school. All people use the same currency. One president unites the country. Islam also joins people together.

The Economy

Indonesia is a developing country. President Suharto has worked hard to improve the economy. Before he became president, millions of people did not have jobs. The country did not grow enough rice for its people. Today most people have jobs. Indonesia now grows so much rice that it exports rice to other countries.

Indonesia earns about one third of its money by exporting oil. The country is a member of the Organization of Petroleum Exporting Countries. OPEC helps decide what the world price of oil should be.

Suharto

Many homes in Indonesia are built on stilts.

Timber is exported from Indonesia to many countries.

A factory in Jakarta

Indonesia has used money from selling oil to develop the country. It has built roads, airports, schools, and factories.

The nation also earns money by exporting other resources. It sells tin, natural gas, timber, and other resources to many countries.

About half of the country's people are farmers. Some people are subsistence farmers. Other farmers work on large plantations. They grow cocoa, coffee, rubber, rice, and spices. Rice and rubber are the most important export crops.

Indonesia has many new industries. Factories make food products, textiles, chemicals, and other goods. But the country still imports most of the factory products it needs.

The people of Indonesia have worked hard to become a united nation. They continue working to improve their standard of living.

Chapter Main Ideas

1. Indonesia is a nation of more than 13,670 islands in the Pacific and Indian oceans. It has a tropical climate.
2. Indonesia has the fourth largest population in the world. It is the world's largest Muslim country.
3. Indonesia earns 30 percent of its money by exporting oil. It is a member of OPEC.

◆ Vocabulary

Match Up   Finish the sentences in Group A with words from Group B. On your paper write the letter of each correct answer.

Group A

1. An _____ is a chain of islands.

2. _____ go back and forth between the islands in Indonesia.

3. The dust from an erupting volcano is called _____.

4. People share information when they _____.

5. Many village houses are built on long poles called _____.

6. A damp climate is _____.

Group B

A. Ferry boats

B. communicate

C. stilts

D. archipelago

E. humid

F. volcanic ash

◆ Read and Remember

Finish Up   Choose a word in dark print that best completes each sentence. Write the word on your paper.

Java   volcanoes   Jakarta   Sumatra
fertile   Equator   Irian Jaya

1. All of Indonesia is along the _____.

2. Lava and volcanic ash make Indonesia's soil _____.

3. The capital and trading center of Indonesia is _____.

4. An island with large amounts of oil is _____.

5. Indonesia has about 60 active _____.

6. Two thirds of Indonesia's people live on the island of _____.

7. The part of New Guinea that belongs to Indonesia is _____.

◆ Think and Apply

Find the Main Idea   On your paper copy the boxes shown below. Then read the five sentences. Choose the main idea and write it on your paper in the main idea box. Then find three sentences that support the main idea. Write them on your paper in the boxes of the main idea chart. There will be one sentence in the group that you will not use.

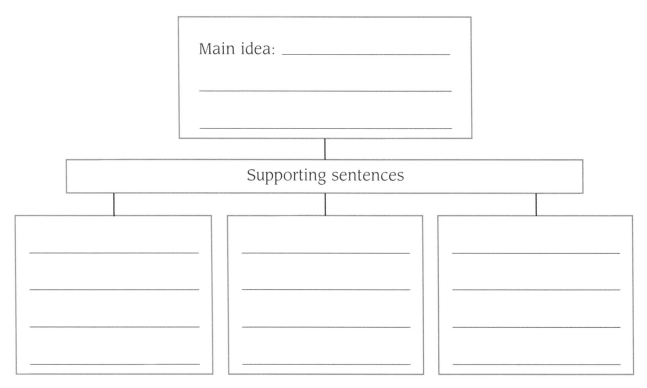

a. Indonesians speak 250 different languages.

b. It is hard for Indonesia to be a united country.

c. It is too expensive for most people to fly from one island to another.

d. All Indonesians use the same currency.

e. Mountains and forests separate people who live on different parts of the same island.

◆ Journal Writing

Write a paragraph in your journal that tells two ways Indonesia is like other countries of South and Southeast Asia. Then tell about two ways it is different from other countries.

Reading a Time Zone Map

There are 24 time zones on a world map. A **time zone map** shows which parts of Earth are in different time periods. As you travel east, you gain one hour for each time zone that you cross. As you travel west, you lose one hour for each time zone you cross. Indonesia has three time zones. Some time zones curve around islands so that all of the people on an island have the same time. When it is 12:00 noon in Sumatra, it is 1:00 P.M. in Borneo, and 2:00 P.M. in Irian Jaya.

Study the time zone map. On your paper write the answer to finish each sentence.

1. If it is 6:00 A.M. in Jakarta, it is _____ in Borneo.

10:00 P.M.   9:00 P.M.   7:00 A.M.

2. If it is 6:00 A.M. in Sumatra, the time in Vietnam is _____.

3:00 P.M.   6:00 A.M.   11:00 P.M.

3. If it is 12:00 noon in Borneo, it is 11:00 A.M. in _____.

Java   Sulawesi   Irian Jaya

4. If it is 12:00 noon in Singapore, it is the same time in _____.

Sumatra   Borneo   Java

5. If it is 2:00 P.M. in Java, it is _____ in Irian Jaya.

4:00 P.M.   9:00 P.M.   11:00 P.M.

6. Sulawesi will have the same time as _____.

Sumatra   Timor   Irian Jaya

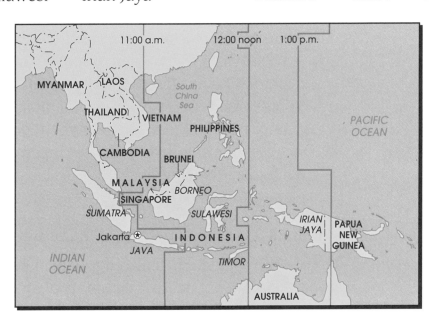

Indonesia: Time Zone Map

CHAPTER 40

Problems Facing South and Southeast Asia

Where Can You Find?

Where can you find a country in Southeast Asia that is a center for world trade?

Think About As You Read

1. How can nations solve the problems of overpopulation and overcrowded cities?
2. How are nations solving the deforestation problem?
3. How has the Green Revolution helped India?

New Words

- sanitation
- Green Revolution
- miracle seeds
- foreign investments

People and Places

- Singapore
- Bangkok, Thailand

The nations of South and Southeast Asia are developing countries. Singapore is the only developed country in this region. These countries must solve five problems in order to help their people live better lives.

Poverty, Overpopulation, and Overcrowded Cities

Poverty is a big problem in most countries of the region. In many countries people earn less than $1,500 a year. Cities like Calcutta, India, have millions of poor people. Most people in the region live in small villages. They work as subsistence farmers. They often do not have extra crops that they can sell to earn money. Governments need to teach farmers modern methods to grow food. Then the farmers can earn more money by selling their crops.

Overpopulation is also a problem. Populations are growing very fast. In India one million children are born each month. The population of this region will

Schoolgirls in Thailand waiting in line to play

Tiny Singapore has proved that countries in this region can have a high standard of living. Singapore has become a busy center for world trade.

After many years of war, Vietnam is becoming a better place to live. Children are going to school and learning to read. The literacy rate is high. Tourists are visiting the country and spending money.

The people of South and Southeast Asia want to solve their problems. Then more people will enjoy a higher standard of living.

Chapter Main Ideas

1. The five big problems in this region are poverty, overpopulation, overcrowded cities, deforestation, and war.
2. Foreign investments are helping some nations build new factories.
3. The Green Revolution has helped India grow more food.

An American tourist shows a picture to these Vietnamese children.

Singapore

Singapore is a rich developed country in Southeast Asia. It is a small country with about 3 million people. The country has one main island called Singapore. There are also about 50 very small islands. The capital and only city is in the south of the main island. The capital is the city of Singapore. There are small towns on other parts of the island. Most of the nation's people live in the capital.

Singapore's location is 70 miles from the Equator. So it has a hot, rainy climate. Singapore is near Malaysia and Indonesia. It is on a trade route between Europe, Africa, and Asia. Ships from many nations use the port of Singapore as they travel from one continent to another. Singapore's harbor is one of the world's busiest ports.

Singapore's factories make many kinds of products. They use modern technology. These products are sent to many parts of the world.

At one time many wild animals lived on the island of Singapore. As Singapore built roads and factories, most of the animals were destroyed.

Singapore is a region of very strict government. It is against the law to spit or to cross the street when the traffic light is red. Newspapers cannot write against the government. People are punished when they break the country's strict laws. Most people in Singapore obey the laws.

SINGAPORE

MALAYSIA

SINGAPORE

⊛ Singapore

Strait

SINGAPORE ISLAND

Singapore

MAP KEY
⊛ Capital city

Write a sentence to answer each question.

1. **Place** What kind of place is Singapore?

2. **Human/Environment Interaction** How did the development of Singapore affect wildlife?

3. **Region** What kind of region is Singapore?

4. **Location** Where is Singapore?

5. **Movement** How does the port of Singapore help trade?

USING WHAT YOU'VE LEARNED

◆ Vocabulary

Find the Meaning Choose the word or words that best complete each sentence. Write your answers on your paper.

1. A city with poor **sanitation** will be ______.

   clean   dirty   modern

2. Singapore has **foreign investments** because people from other nations have started ______.

   businesses and factories   tariffs and taxes   wars and farms

3. The **Green Revolution** helps ______.

   lawyers   painters   farmers

4. **Miracle seeds** produce ______.

   larger crops   smaller crops   earlier crops

◆ Read and Remember

Complete the Chart Copy the chart shown below on your paper. Use facts from the chapter to complete the chart. You can read the chapter again to find facts you do not remember.

Problems of South and Southeast Asia

	What Is the Problem?	How Does the Problem Hurt the Region?	What Is One Way the Problem Is Being Solved?
1.			
2.			
3.			
4.			
5.			

Write the Answer   Write one or more sentences to answer each question.

1. How many children are born in India each month?

2. Which cities in this region have more than 10 million people?

3. Where in the region have religious groups fought each other?

4. When did the Vietnam War end?

5. How have miracle seeds helped India?

6. What has helped Singapore become a developed country?

◆ Think and Apply

Cause and Effect   Number your paper from 1 to 6. Write sentences on your paper by matching each cause on the left with an effect on the right.

Cause

1. Overpopulation is a big problem, so _____.

2. Many forests have been chopped down, so _____.

3. The people of Bangladesh wanted to be independent from Pakistan, so _____.

4. The Green Revolution helped India, so _____.

5. Foreign businesses are starting new factories, so _____.

6. Vietnam now has peace, so _____.

Effect

A. they fought a war in 1971

B. tourists are visiting the country again

C. Indian TV and radio programs tell people to have smaller families

D. there is a lot of erosion in the region

E. some countries in the region are becoming more developed

F. the country now exports rice and wheat

◆ Journal Writing

Write a paragraph about two problems in South and Southeast Asia. Explain what the problems are. Then tell how nations can try to solve them.

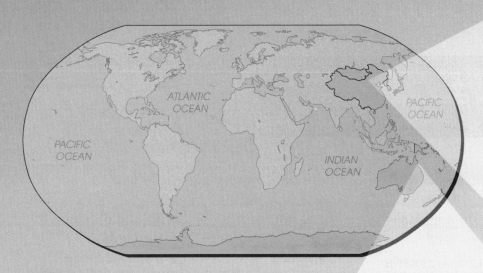

The Great Wall of China

DID YOU KNOW?

▲ No one has ever been able to count all of the islands in the Pacific Ocean. There may be as many as 30,000 islands.

▲ The world's longest railroad tunnel is in Japan. It connects the islands of Honshu and Hokkaido.

▲ More people speak Chinese than any other language.

▲ China shares its border with 14 nations. No other country has borders that touch as many nations.

▲ There are more sheep than people in Australia and New Zealand.

The Great Barrier Reef

WRITE A TRAVELOGUE

Look at the photographs in Unit 8. Choose one country you would like to visit because it reminds you of the United States. Choose a second country you would like to visit because it looks different from the United States. In your travelogue write about why you want to visit each country. After reading Unit 8, write about three more countries you would like to visit. Tell why each country is a special place.

THEME: PLACE

CHAPTER 41

Looking at East Asia and the Pacific

Where Can You Find?

Where can you find a country that is also a continent?

Think About As You Read

1. What factory products do Americans buy from this region?
2. What are the landforms and climates of East Asia and the Pacific?
3. How do natural disasters hurt this region?

New Words

- mainland
- temperate zones
- inland
- natural disasters
- tsunamis
- free enterprise

People and Places

- Oceania
- New Zealand
- China
- Hong Kong
- North Korea
- South Korea
- Taiwan
- Hong Kong City
- Kowloon
- Special Administrative Region of China (S.A.R.)

American stores are filled with many products from the countries of East Asia and the Pacific. There are cameras from Japan and bicycles from Taiwan. There are cotton shirts from China and wool sweaters from Australia. This is a region of trade and industry.

The Region of East Asia and the Pacific

East Asia and the Pacific is a huge region. It has four parts. The Equator divides this region. Three parts are south of the Equator. The first part is Australia. It is a country that is also a continent.

The second part is Oceania. Oceania has the islands of the Pacific Ocean. There are about 25,000 islands in Oceania. New Zealand is the only developed country in Oceania. Oceania and Australia are not crowded. There is room for many more people.

Antarctica is the third part of the region. It is a frozen continent at the South Pole.

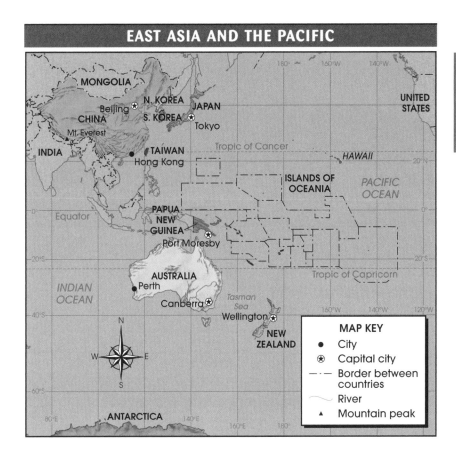

East Asia and the Pacific is a huge region divided by the Equator. What two countries north of the Equator share a peninsula?

The fourth part of this region is north of the Equator. The countries of East Asia cover this part of the region. China, Japan, and Hong Kong are in East Asia. North Korea, South Korea, and Taiwan are also in East Asia. They have most of the region's cities, people, and factories. They are all very crowded. China is the largest country in East Asia. It has more people than any other country in the world.

Hong Kong is a special region in East Asia. It is a busy port and a center of world trade. Hong Kong includes a peninsula off the coast of southeastern China and many small islands. Hong Kong City and Kowloon are its largest cities. Hong Kong was part of China for more than 1,000 years. From the mid-1800s to 1997, the British ruled Hong Kong. Since July 1997, Hong Kong has been ruled by China again. Hong Kong is called the Special Administrative Region of China, or S.A.R. About 6 million people live in this small region.

The busy port of Hong Kong

Landforms and Climate

East Asia includes land on the Asian continent called the **mainland**. East Asia also includes the

An island in the Pacific

island countries of Japan and Taiwan. These islands are near the mainland.

This region has many landforms. Mountains cover most of East Asia. There are plains near the coasts and near the rivers. Most East Asians live on these crowded plains. Many islands in the Pacific are covered with mountains. Most of Australia is covered with plains and plateaus.

The region has many different climates. The climates to the north and south of the tropics are called the **temperate zones**. The climate in these zones is not too hot or too cold. Japan, Korea, and most of China are in the northern temperate zone. New Zealand and southern Australia are in the temperate zone south of the Equator.

Monsoons bring summer rains to the coasts of East Asia. **Inland** areas, or places that are far from the ocean, are drier. Parts of western China have a desert climate. Most of East Asia has cold, dry winters.

Winter in northern Japan

Most of the islands of Oceania have a warm climate. New Zealand and some other islands get a lot of rain. Other islands get only a few inches of rain a year. Northern Australia has a warm, dry climate. The climate in southern Australia is cooler and rainier.

Natural Disasters

Natural disasters are terrible accidents caused by nature. Many natural disasters happen in East Asia and the Pacific. Japan, Taiwan, and many Pacific islands have volcanoes that sometimes erupt. Erupting volcanoes destroy homes and sometimes kill people.

This part of the world also has many earthquakes. In an earthquake the ground shakes and the land cracks. Earthquakes destroy homes, buildings, and bridges. Many people have been killed during strong earthquakes. This region has more earthquakes and volcanoes than any other part of the world.

Typhoons also destroy homes, farms, and buildings in this region. These dangerous tropical storms bring high winds and heavy rains.

Tsunamis are huge, dangerous waves. They are caused by underwater earthquakes. They crash into the coasts of East Asia and the Pacific Islands.

An earthquake destroyed these buildings in Japan.

Governments and Economies

There are different kinds of governments in East Asia and the Pacific. Australia, New Zealand, and Japan are democracies. China and North Korea have Communist governments. Dictators lead these governments. The people have little freedom.

China and North Korea have command economies. In a command economy, the government owns all of the farms, factories, and businesses. The government decides salaries and prices. In the 1980s China's government began to allow some **free enterprise**.

Japan, South Korea, and Hong Kong have free market economies. People can own and control their own businesses. They decide salaries and prices. This is also called free enterprise. But now China may have more control over Hong Kong's businesses.

Japan, South Korea, Hong Kong, and Taiwan have many factories and industries. They have strong economies. They manufacture cameras, computers, cars, and hundreds of other products. They are important centers for trade. They earn billions of dollars by exporting their products to all parts of the world. Their people enjoy a high standard of living. Australia and New Zealand also have high standards of living.

Tsunamis cause damage to islands in the Pacific.

China is still a developing nation. Most of the people are farmers. But the country now has many kinds of factories. China exports factory products to all parts of the world.

The countries of East Asia are all very crowded. Which country has the most cities with more than 2 million people?

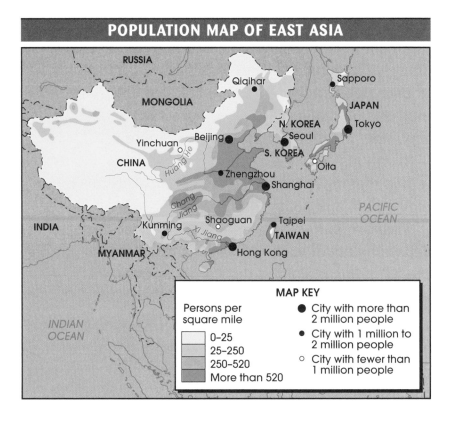

POPULATION MAP OF EAST ASIA

RUSSIA

MONGOLIA

Qiqihar

Sapporo

JAPAN

N. KOREA
Seoul

Tokyo

Yinchuan
Beijing

CHINA

Huang He

S. KOREA

Oita

Zhengzhou

Shanghai

PACIFIC
OCEAN

Chang
Jiang

INDIA

Kunming
Xi Jiang

Shaoguan

Taipei

TAIWAN

MYANMAR

Hong Kong

INDIAN
OCEAN

MAP KEY

Persons per
square mile

☐ 0–25
☐ 25–250
☐ 250–520
☐ More than 520

● City with more than
2 million people

• City with 1 million to
2 million people

○ City with fewer than
1 million people

A Buddhist shrine in Japan

A Region of Many Cultures

The people of this region belong to many ethnic groups. They speak many different languages.

Most people in Australia and New Zealand are Christians. Buddhism is an important religion in East Asia. Communist leaders in China and North Korea discourage people from practicing any religion.

More people live in East Asia and the Pacific than in any other part of the world. Americans use products from this region every day. As you read Unit 8, find out why this region is important in today's world.

Chapter Main Ideas

1. East Asia has crowded countries. Most of the people live on plains near the coasts and rivers. Australia and the Pacific Islands are not crowded.
2. Japan, South Korea, Hong Kong, and Taiwan have many industries. They export products to many countries.
3. China has more than one billion people. Its Communist government now allows people to own businesses and factories.

◆ Vocabulary

Analogies   Choose the word or words in dark print that best complete the sentences. Write the word or words on your paper.

> **tsunami**   **temperate zones**   **free enterprise**
> **inland**   **natural disaster**   **mainland**

1. Japan is to island as China is to _____.

2. Hot climate is to tropics as mild climate is to _____.

3. Coastal is to near the coast as _____ is to away from the coast.

4. Deforestation is to a problem made by people as earthquake is to a _____.

5. Typhoon is to storm as _____ is to wave.

6. Communism is to command economy as _____ is to free market economy.

◆ Read and Remember

Finish Up   Choose the word or words in dark print that best complete each sentence. Write the word or words on your paper.

> **temperate zones**   **East Asia**   **Communist**   **Hong Kong**
> **summer**   **Oceania**   **dictators**

1. There are about 25,000 Pacific islands in an area called _____.

2. Most of this region's people, factories, and cities are in _____.

3. Japan and New Zealand are in the _____ because they are north and south of the tropics.

4. Monsoons bring _____ rain to the coast of East Asia.

5. Some countries in this region are ruled by leaders called _____.

6. China and North Korea have _____ governments.

7. _____ is called the Special Administrative Region of China, or S.A.R.

Write the Answer   Write one or more sentences to answer each question.

1. What products are made in East Asia?

2. What countries are in the temperate zone?

3. What are the developed countries of this region?

4. What is the only developed country in Oceania?

5. What natural disasters happen in East Asia and the Pacific?

6. What different kinds of governments are in this region?

7. When did China start to control Hong Kong?

◆ Think and Apply

Categories   Find the best title for each group from the words in dark print. Write the title on your paper.

High Standard of Living   **The Pacific Region**   **Australia**
Natural Disasters   **East Asia**   **China**

1. Antarctica
Australia
Oceania

2. Taiwan and Hong Kong
North and South Korea
China and Japan

3. typhoons
earthquakes and volcanoes
tsunamis

4. Australia and New Zealand
Japan and South Korea
Hong Kong and Taiwan

5. more than one billion people
Communist government
allows some free enterprise

6. a continent and a country
plains and plateaus
democracy

◆ Journal Writing

East Asia and the Pacific is a region of trade and industry. Look at the car, television, radio, camera, or other products your family or someone you know may own. Which products are from East Asia and the Pacific? Write a paragraph in your journal that tells where many of these products were produced.

CHAPTER 42

China: Asia's Largest Nation

Think About As You Read

1. What happened at Tiananmen Square?
2. How did Deng Xiaoping help China's economy?
3. How has the standard of living in China improved?

New Words

- protest
- demonstrations
- collectives
- contracts
- profits
- dikes
- characters
- herbs
- acupuncture

People and Places

- Huang He (Yellow River)
- the Great Wall of China
- Tiananmen Square
- Beijing
- Deng Xiaoping
- Xi Jiang (West River)
- Chang Jiang (Yangtze River)
- Shanghai
- Confucius

More than one billion people live in China. It has more people than any other country in the world. It is only a little larger than the United States. But China struggles to feed four times more people than the United States.

History, Government, and Protests at Tiananmen Square

China is a very old country. Thousands of years ago, people settled on fertile plains around the Huang He River. Later they settled in other parts of China.

More than 2,000 years ago, the Chinese began building the Great Wall of China. They built the wall across northern China. It was built to stop enemies from attacking China. When it was finished, the wall was more than 1,500 miles long. Many tourists still visit the Great Wall each year.

For thousands of years, most Chinese were hungry and poor. Many people believed communism would solve China's problems. In the early 1900s, Chinese

China has many kinds of landforms. What mountain chain is located in the southwestern part of China?

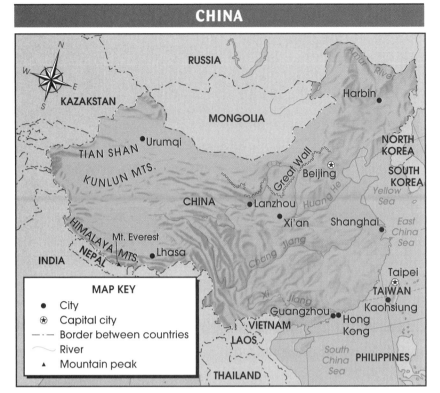

CHINA

MAP KEY
- • City
- ⊛ Capital city
- –·– Border between countries
- ⌒ River
- ▲ Mountain peak

China's flag

Student protestors at Tiananmen Square

Communists began fighting to control the government. Since 1949 Communists have ruled China.

China's non-Communist leaders fled China in 1949. They escaped to the nearby island of Taiwan. In Taiwan they started a non-Communist government. Today Taiwan is a developed country with many industries. Taiwan has a much higher standard of living than China.

The Chinese Communist party has ruled China's government since 1949. People have little freedom. There is no freedom of speech. People cannot **protest** against the government.

In 1989 Chinese students held **demonstrations** in Tiananmen Square to show they wanted more freedom. This happened in the capital, Beijing. China's leader, Deng Xiaoping, sent Chinese soldiers to attack the protesting students. Many students were killed. There is still little freedom in China today.

Economy and Standard of Living

China has a command economy. For many years the government owned all farms, factories, and businesses. But China still had millions of hungry people.

In the 1980s Deng Xiaoping began to improve the economy. He allowed foreign countries to invest their money in Chinese factories. He allowed trade with many countries. Deng bought new machines and technology from other countries. He started new factories in villages. Deng allowed people to own their own small restaurants and businesses.

Deng Xiaoping wanted farmers to grow more food. He knew farmers were not happy working on the huge government-owned farms called **collectives**.

Deng decided that the government would give **contracts** to all farmers. These contracts told farmers how much to grow and how much to sell to the government. Farmers could keep all extra crops for themselves. They could sell these crops and keep the **profits**. So the farmers worked harder to grow much more food. China now has enough food for its huge population. It exports rice to other countries.

China is still a poor country. But most Chinese enjoy a higher standard of living today than they did before 1949. Most children now go to school. Most people receive medical care. Most people now own a bicycle, a radio, and a sewing machine. The standard of living is a little higher in cities. But people who live in the cities live in tiny apartments. Few people in China own cars. Most people travel on bicycles. There are more than 200 million bicycles in China.

These men own their own grocery store in northern China.

A Chinese village

Mountains and deserts cover much of western China.

China's Landforms, Climate, and Resources

China has many kinds of landforms and climates. Less than 15 percent of all Chinese land can be farmed. Much of China's land in the west is covered with mountains and plateaus. Only a small part of the population lives there. The tall Himalaya Mountains are in the southwest. Large deserts also cover part of the west. Western China has a cold, dry climate.

Most of the country's cities, farms, and people are in eastern China. The east has a long coast. Plains, hills, and low mountains cover the east. People live on fertile plains near the coast and the rivers.

Eastern China has three long rivers. These rivers are surrounded by fertile plains. The Huang He (Yellow River) is in the north. The Xi Jiang (West River) is in southern China. China's longest river is the Chang Jiang (Yangtze River). It is the third longest river in the world. Ships travel on the Chang Jiang for hundreds of miles.

The Huang He has had many floods. These floods have destroyed farms and villages. But silt in the flood waters has made the river valley fertile. To control the floods, the Chinese have built dams and **dikes**, or special walls, on the river.

Monsoons bring rain to eastern China in the summer. The southeast is warmer and wetter than other parts of China. The Chinese grow rice there. The climate is cooler in the north. So northern farmers grow wheat instead of rice.

A ship on the Chang Jiang

China has many natural resources. It has oil, coal, and iron. It uses waterpower from its rivers to make electricity. China has gold, silver, and tin. Uranium and bauxite are other important minerals.

Cities, Language, and Culture

Today about one fourth of China's people live in cities. There are 30 cities that have more than one million people. Shanghai is China's largest city. It has more than 12 million people. It is an important port. It has many factories. Beijing is China's capital. It is near iron and coal mines. This city has large steel factories.

Shanghai

The Chinese language does not have an alphabet. Instead the Chinese write with little pictures called **characters**. Each character shows a word or an idea. To read a Chinese newspaper, a person must know at least 3,000 characters.

The Chinese have followed the teachings of their great teacher Confucius for more than 2,500 years. Confucius taught people to respect parents, teachers, and government leaders. He also said people must be honest and kind.

For thousands of years, the Chinese have made medicines from plants called **herbs**. They also invented a system called **acupuncture**. Acupuncture doctors push small needles into different parts of the body. These needles treat disease and stop pain.

Looking at the Future

China's leaders believe the country will not have enough food if its population continues to grow. So China's laws now allow couples to have only one child. Some couples are allowed to have two children. Couples are punished if they have more children.

A statue of Confucius

In July 1997 China began to rule the small region of Hong Kong. Hong Kong had been ruled by Great Britain since the mid-1800s. Hong Kong has a much higher standard of living than China.

China's Communist leaders want Hong Kong to remain a rich industrial region. They want Hong Kong to share the money it earns with China. So most

Making medicine from herbs

people believe Hong Kong will continue to have a free market economy. But the people of Hong Kong may have less freedom. No one knows how China's Communist government will change Hong Kong.

The world is watching China. Can China continue to grow enough food for its huge population? Will people be allowed to enjoy more freedom? China will always be an important country.

Chapter Main Ideas

1. Most of China's people live in the east. There are fertile plains near the coast and around three long rivers.
2. Since 1949 China has been a Communist country. The people have little freedom.
3. Deng Xiaoping improved China's economy. Farmers grow much more food. There is more industry.

BIOGRAPHY

Deng Xiaoping (1904–1997)

Deng Xiaoping became China's powerful leader in 1977. At the time he was 72 years old. China was a very poor country.

Deng decided that free enterprise would help China's economy. Most Communists were against free enterprise. But Deng wanted to help China's economy. He allowed farmers and business owners to make profits. Many new factories were started. Deng allowed foreign countries to start factories in China. He allowed the United States and other countries to have more trade with China. China's economy grew stronger.

Deng did not believe in democracy. He sent the Chinese Army to attack the student protesters at Tiananmen Square.

Deng died when he was 92 years old. He had helped China become a more modern nation.

Journal Writing
Write a paragraph in your journal that tells how Deng Xiaoping helped and hurt China.

◆ Vocabulary

Finish Up   Choose the word in dark print that best completes each sentence. Write the word on your paper.

collectives	**contracts**	**demonstrations**	**dikes**
profits	**protest**	**characters**	**acupuncture**

1. The symbols that are used in Chinese writing are called _____.

2. Walls that are built to stop rivers from flooding are _____.

3. Huge farms that are owned by the government are called _____.

4. When you speak out against the government, you _____.

5. Large meetings of people who want changes are _____.

6. The money people earn from selling their products is called _____.

7. Written agreements between a person and the government or between two or more people are called _____.

8. The Chinese invented a system of medicine called _____.

◆ Read and Remember

Write the Answer   Write one or more sentences to answer each question.

1. Where did the early Chinese people settle?

2. What happened in China in 1949?

3. What kind of government does Taiwan have?

4. What happened at Tiananmen Square in 1989?

5. How do most Chinese travel?

6. How many children does the government allow a couple to have?

7. What are China's three long rivers?

8. Where do most of China's people live?

9. What is China's capital city?

10. What happened in July 1997?

Matching Each item in Group A tells about an item in Group B. Write the correct letter next to each number on your paper.

Group A

1. This leader helped China become a more modern nation.

2. This is China's longest river.

3. These bring rain to eastern China in the summer.

4. China's non-Communist leaders fled to this island in 1949.

5. This was built 2,000 years ago to stop China's enemies.

Group B

A. Chang Jiang

B. Taiwan

C. Deng Xiaoping

D. The Great Wall of China

E. monsoons

◆ Think and Apply

Sequencing Number your paper from 1 to 5. Write the sentences on your paper to show the correct order.

Non-Communists escaped to the island of Taiwan.

Chinese Communists fought to win control of the government.

Deng Xiaoping sent soldiers to attack protesting students at Tiananmen Square.

China took control of Hong Kong.

China became a Communist country.

◆ Journal Writing

The laws of China allow most couples to have only one child. Write a paragraph in your journal that tells why China has this law. Tell whether you think the law will help China.

Reviewing Latitude and Longitude

Every place on Earth has its own latitude and longitude. Shanghai, China, has a latitude of 31°N and a longitude of 121°E. We say that the latitude and longitude of Shanghai is 31°N, 121°E.

Look at the map of China below. Then finish each sentence in Group A with an answer from Group B. Write the letter of the correct answer on your paper.

Group A

1. The latitude of Beijing is _____.

2. The longitude of Guangzhou is _____.

3. The city of Urumqi has a latitude and longitude of _____.

4. The city of _____ has a latitude and longitude of 29°N, 91°E.

5. The city of _____ has a latitude and longitude of 46°N, 127°E.

Group B

A. 40°N

B. 113°E

C. Lhasa

D. Harbin

E. 44°N, 87°E

China: Latitude and Longitude

CHAPTER 43

Japan: Asia's Industrial Leader

Where Can You Find?
Where can you find the palace of the emperor?

Think About As You Read

1. Why is Japan a crowded country?
2. What kinds of industries are in Japan?
3. Why does Japan have a favorable balance of trade?

New Words

- Shinto
- kimonos
- bonsai
- robots
- terraces
- dormant

People and Places

- Honshu
- Mount Fuji
- Kobe
- Sea of Japan
- Tokyo
- Sapporo
- Hokkaido

Japan is a crowded country with few natural resources. There is little farmland. Most people would expect Japan to be a poor country. But Japan is the richest country in Asia. It is Asia's industrial leader.

Japan's Landforms, Climate, and Resources

Japan is an archipelago country. It has four large islands. There are also thousands of small islands. Honshu is the largest island. Most Japanese people live on Honshu.

Hills and mountains cover most of Japan. Thick forests cover Japan's mountains. About 60 of the mountains are volcanoes that sometimes erupt. Japan's tallest mountain is a beautiful volcano called Mount Fuji.

Narrow coastal plains cover about one fifth of Japan. Most people live on these plains. Japan is a very crowded country because many people live on this small amount of land.

Japan has more earthquakes than any other nation. It has more than 1,000 earthquakes each year. Most of

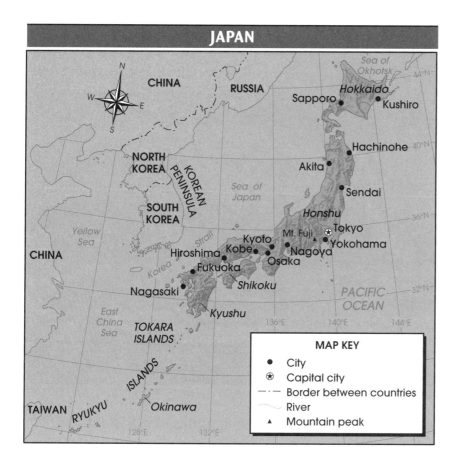

JAPAN

Japan is an archipelago country with four large islands and many small ones. What is the name of Japan's largest island?

Japan's flag

them are so small that no one feels them. But some cause great damage. In 1995 an earthquake in the city of Kobe killed more than 5,000 people.

The seas are important to Japan. The Sea of Japan is to the west of Japan. The Pacific Ocean is to the east. No place in the country is more than 100 miles from the sea. The Japanese use the seas for shipping and trading. Fishing is an important industry. The Japanese eat a lot of fish.

Japan's climate changes as you move from south to north. Islands in the south have hot summers and mild winters. The northern islands have cold winters and cool summers. But the island of Honshu has warm, humid summers and cold winters.

Mount Fuji is a beautiful volcano.

Japan has few resources. It gets wood from its forests. It creates some hydroelectric power from its rivers. Japan must import coal, oil, and raw materials.

People, Culture, and Government

About 126 million people live in Japan. They speak the Japanese language. They use thousands

This man grows bonsai trees.

of characters to write their language. It is a difficult language to read. But everyone in Japan knows how to read and write. The Japanese people spend many years in school.

Many people in Japan practice both the **Shinto** and Buddhist religions. People who believe in the Shinto religion pray to many gods. These gods are found in rivers, mountains, trees, and other forms of nature. The country has thousands of Shinto and Buddhist temples.

Japan is a very modern country, but the Japanese people love their old traditions. People bow to each other to say hello and goodbye. They show great respect for parents and for older people. They remove their shoes before they enter a home. They cover the floors of their homes with straw mats. People sometimes wear beautiful robes called **kimonos**. Some people enjoy planting small beautiful gardens. They grow tiny trees called **bonsai** trees in flower pots.

The government of Japan is a democracy and a constitutional monarchy. It has a parliament and a prime minister. It also has an emperor. The emperor has no power in the government.

A Shinto shrine

Japan: An Industrial Giant

Japan has a strong economy. Japan makes more factory products than all of the nations in Western Europe. In some years Japan produces more cars than the United States. The Japanese make excellent televisions, cars, cameras, computers, and other products. Their factories have the newest technology. Factories often use machines called **robots** to do many jobs.

Trade has helped Japan become a rich country. Japan imports raw materials for its factories. It exports huge amounts of factory products. Japan limits the number of products it buys from the United States and other countries. So Japan has a favorable balance of trade. It exports much more than it imports.

About one third of the Japanese people work for large companies. These companies make their workers feel like they are part of a large family. They treat their workers well. People work hard and receive good salaries. Many people work for one company until they are ready to retire.

Robots do many jobs in Japan. This robot is repairing power lines.

Farms, Cities, and Standard of Living

Japan has many mountains, so there is little farmland and most farms are small. Only a small part of the population works at farming. The Japanese are excellent farmers. They build **terraces** into the sides of hills so

The Japanese grow rice on terraces built into the sides of hills.

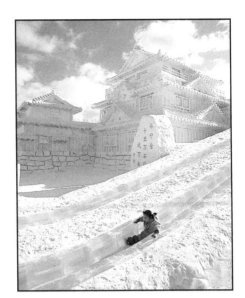
Sapporo's ice festival

City workers helping people into a Tokyo subway

they can grow more crops. Terraces are large, flat areas of land for planting crops. The Japanese use fertilizers, good seeds, and modern machines. Farmers grow about two thirds of the food the country needs. Rice is the most important crop. Japan must import some of its food.

About three fourths of Japan's people live in cities. Most big cities are on the coast of the island of Honshu. Tokyo is Japan's capital and largest city. It is also a busy port. Tokyo is one of the largest cities in the world. It is also one of the most crowded. Most people travel around Tokyo by subway. Special city workers push people into the crowded trains so the doors will close. Tokyo is Japan's main business and arts center. It is also the home of the emperor's palace. People are allowed to visit the gardens around this famous palace.

Sapporo is the only large city on the northern island of Hokkaido. Every winter people visit Sapporo's ice festival. Visitors see temples, buildings, and statues that are carved out of ice. People can travel from Tokyo to Sapporo in very fast bullet trains. These trains go through a long tunnel that connects the islands of Honshu and Hokkaido.

Most people in Japan are part of the middle class. They have a good life. But Japan has two big problems. It is too crowded. There is a lot of pollution. Cars and factories are making the air and water dirty.

Look around your home. Look at the cars in the street. You will see many products from Japan. Japan is Asia's industrial leader.

Chapter Main Ideas

1. Japan has few natural resources. It uses imported raw materials in its factories.
2. Most of Japan is covered with mountains. Most people live on narrow coastal plains.
3. Japan is the richest country in Asia. It imports large amounts of factory products. Most people belong to the middle class.

The Ring of Fire

The Ring of Fire is a chain of volcanoes that surrounds the Pacific Ocean. Alaska, Hawaii, California, and western South America are in the eastern part of the Ring of Fire. Japan, New Zealand, Indonesia, and other Pacific islands are in the western part. Countries in the Ring of Fire have more earthquakes and volcanoes than countries in other regions.

Volcanoes and earthquakes occur because of changes that take place deep inside the earth. Earth is covered with 30 large, rocky sheets called plates. These plates are under the oceans and continents. Sometimes the plates crash into each other. Sometimes the plates pull apart. These movements cause earthquakes and volcanoes.

Scientists never know when a volcano will erupt. Beautiful Mount Fuji erupted in 1707. Since then it has been **dormant**, or quiet. It is possible Mount Fuji will one day erupt again.

A terrible earthquake destroyed part of Tokyo in 1923. More than 100,000 people died. The Japanese rebuilt their city. But they know it can be destroyed again by another earthquake. Millions of people who live in the Ring of Fire know there can be another natural disaster at any time.

THE RING OF FIRE

ASIA
NORTH AMERICA
PACIFIC OCEAN
AUSTRALIA
SOUTH AMERICA

MAP KEY
▲ Major volcano

Write a sentence to answer each question.

1. **Human Environment/Interaction**   How did the 1923 earthquake affect Tokyo?

2. **Movement**   What can happen when there is movement of Earth's plates?

3. **Place**   What kind of place is Mount Fuji?

4. **Location**   Where is the Ring of Fire?

5. **Region**   Why is this region called the Ring of Fire?

◆ Vocabulary

Match Up   Finish the sentences in Group A with words from Group B. Write the letter of each correct answer on your paper.

Group A

1. One of the main religions of Japan is _____.

2. Japanese robes are called _____.

3. A small tree in a flower pot is a _____.

4. Machines that are made to work like people are called _____.

5. Large, flat areas of land built into a mountain for growing crops are called _____.

6. When a volcano is not active, it is _____.

Group B

A. bonsai

B. terraces

C. Shinto

D. robots

E. dormant

F. kimonos

◆ Read and Remember

Finish the Paragraph   Number your paper from 1 to 7. Use the words in dark print to finish the paragraph below. Write the words you choose on your paper.

archipelago	constitutional monarchy	pollution	raw materials
forests	Honshu	Tokyo	

Japan is an ___1___ country with many islands. The largest island is ___2___. Most Japanese people live on this island. Most of Japan is covered with mountains and ___3___. Some of the mountains are volcanoes. With so many mountains, there is little farmland. Japan also has few natural resources. So, Japan imports large amounts of ___4___ for its factories. Japan's government is a ___5___ because the emperor has no power. The emperor lives in ___6___, the country's capital. One of Japan's biggest problems is ___7___.

Drawing Conclusions   Read each pair of sentences. Then look in the box for the conclusion you might make. Write the letter of the conclusion on your paper.

1. Most of Japan is covered with mountains.
There are narrow plains near the coast.

Conclusion: _____

2. Japan has a large fishing industry.
Japan uses the seas for shipping and trading.

Conclusion: _____

3. Japan must import all of its oil and coal.
Japan must import raw materials for industry.

Conclusion: _____

4. The Japanese bow when they say hello and goodbye.
The Japanese remove their shoes before going into a home.

Conclusion: _____

5. Japan exports more goods than it imports.
Japan limits the amount of goods it buys from other countries.

Conclusion: _____

Conclusions
A. The seas are important to Japan.
B. Japan does not have much farmland.
C. Japan has a favorable balance of trade.
D. Japan has few natural resources.
E. Traditions are important in Japan.

◆ **Journal Writing**

Japan is a crowded country with few resources. Write a paragraph in your journal that tells how Japan became a rich industrial country.

The **bar graph** on this page shows the population of the eight largest cities in Japan. The key shows the islands where the cities are located. Look at all the bars on the graph. Then write the answer to each question on your paper.

1. What is the population of Tokyo?

2. Which cities are not on Honshu?

3. Which two cities have the same population?

4. What city has the second largest population?

5. Which city has the smallest population?

6. How many people live in Osaka?

7. How many people live in Fukuoka?

8. How many cities have more than 3 million people?

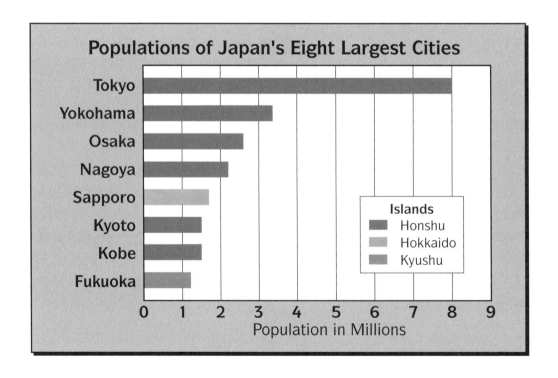

CHAPTER 44

North and South Korea: A Divided Land

Where Can You Find?
Where can you find a strip of land that separates North and South Korea?

Think About As You Read

1. How did Korea become two countries?
2. How are the two Koreas alike and different?
3. How has South Korea changed since 1953?

New Words

◆ Confucianism
◆ allies
◆ Demilitarized Zone

People and Places

◆ Korean Peninsula
◆ Pyongyang
◆ Seoul

If you lived in South Korea, you would never see your family members who live in North Korea. People in the two countries cannot send letters to each other. They cannot make phone calls to each other. As you read this chapter, find out why Korea is a divided land.

Landforms and Climate of the Korean Peninsula

North and South Korea share the Korean Peninsula. China and Russia are Korea's northern neighbors. The Sea of Japan is between Korea and Japan.

Many parts of the peninsula are covered with forests. Mountains cover most of the Korean Peninsula. So most land cannot be farmed. The east coast is mountainous. Plains cover the western coast and part of the south. Most people live on these plains. Most Korean farmland is on these plains.

Most of Korea has warm summers and cold winters. Korea receives rain throughout the year. But monsoon winds bring heavy rain from June to August.

North Korea is separated from South Korea by a strip of land at the 38th line of latitude. What is this strip of land called?

NORTH KOREA AND SOUTH KOREA

MAP KEY
- • City
- ⊛ Capital city
- —·— Border between countries
- ∿ River
- ▲ Mountain peak

RUSSIA
CHINA
Mt. Paektu
Chongjin
Yalu River
NANGNIM MOUNTAINS
HAMYONG MTS.
NORTH KOREA
Sinuiju
Hamhung
Sea of Japan
Pyongyang ⊛
Nampo
Demilitarized Zone
38th parallel
CHINA
N
W   E
S
Seoul ⊛   SOUTH KOREA
TAEBAEK MOUNTAINS
Taejon
SOBAEK MTS.
Chonju
Taegu
Yellow Sea
Kwangju
Pusan
Korea Strait
122°E   124°E   126°E   Cheju   128°E   JAPAN
42°N   40°N   38°N   36°N

North Korea's Flag

South Korea's Flag

All Koreans share a very old culture.

Culture and History

For hundreds of years, all Koreans shared the same culture. The people of North and South Korea speak the Korean language. They write with the Korean alphabet. Education is an important part of Korea's culture. Koreans also have great respect for parents and older people.

Korea's history began long ago. About 4,500 years ago, Koreans started a small country near Pyongyang. Pyongyang is now the capital of North Korea. Later, different Korean tribes ruled different parts of the peninsula. At times China conquered and ruled part of the peninsula. The Chinese brought Buddhism and **Confucianism** to Korea. Confucianism is a way of living based on the teachings of Confucius.

North and South Korea were one country for most of the nation's history. Japan won control of Korea in 1910. It ruled Korea until World War II ended in 1945.

In 1948 Korea was divided at the 38th parallel, or line of latitude. North Korea had a Communist

354

government. The Soviet Union and China were its **allies**, or friends. South Korea had a non-Communist government. The United States was South Korea's ally.

In 1950, North Korea invaded South Korea. This was the start of the Korean War. North Korea's goal was to unite the entire country under a Communist government. The United Nations sent soldiers to help South Korea. Most of the soldiers were Americans. China helped North Korea. The war ended in 1953. South Korea remained an independent country.

But North and South Korea continue to be separated. A strip of land that is two and a half miles wide separates the two Koreas at the 38th parallel. This strip of land is called the **Demilitarized Zone**, or DMZ. No one lives on this land. Soldiers are not allowed on this land. Soldiers from both sides guard the borders of their countries at the DMZ. The United States has thousands of soldiers in South Korea. They help guard South Korea's border.

In 1972 and 1991 North and South Korea tried to have peace talks. But the two Koreas are still enemies.

South Korean troops guarding the border at the DMZ

Looking at North Korea

North Korea has 24 million people. More than half of the people live in cities or towns.

North Korea is more mountainous than South Korea. It has many natural resources. It has plenty of coal, copper, iron, and lead.

North Korea is a poor country. Its Communist government owns all of the country's factories and businesses. Factories do not make enough consumer goods. All farmers work on large collective farms. The farmers do not grow enough food. In 1995, 1996, and 1997 there were serious food shortages.

There is no freedom in North Korea. People cannot speak or write against the government. The government controls the newspapers and the radio and television stations.

North Korea is less developed than South Korea. Its factories do not have as much modern technology. It has only one large city. That city, Pyongyang, has more

Pyongyang is the only large city in North Korea.

Seoul, South Korea

than 2 million people. Few people own their own cars in North Korea. Many people in the cities ride bicycles.

Looking at South Korea

South Korea was a poor country when the Korean War ended. Most people were farmers. Since then South Korea has built many factories and industries. It is now a rich industrial country. People enjoy a higher standard of living than people in North Korea.

South Korea has few natural resources. Like Japan, it must import raw materials. South Korea uses imported raw materials to make many products. Today South Korean factories produce clothes, cars, ships, steel, and electronic products. Each year it earns more and more money from exporting its products.

Today South Korea has about 45 million people. South Korea is smaller than North Korea, but its population is larger. So it is more crowded.

This factory in South Korea produces cars for export.

About three fourths of South Koreans live in large cities. Many people move to the cities to get jobs. Some large cities are very crowded. Seoul, the capital, is a modern city with more than 10 million people.

People enjoy life in South Korea. Farmers grow enough food. The middle class is growing. People are earning higher salaries. Many people own cars.

South Koreans are also enjoying more freedom. For years after the Korean War ended, different groups tried to rule the country. Sometimes people could not criticize the government. But in 1987 the country wrote a new constitution. It allows more democracy and freedom.

No one knows what the future of the two Koreas will be. Many Koreans hope that one day their country will be united again.

Chapter Main Ideas

1. North Korea is a poor Communist country. South Korea is a rich non-Communist country.
2. The Korean War began in 1950 when North Korea invaded South Korea. The war ended in 1953.
3. The Demilitarized Zone separates North and South Korea at the 38th parallel.

◆ Vocabulary

Find the Meaning   Write the word or words that best complete each sentence. Write your answers on your paper.

1. The **Demilitarized Zone** in Korea is a region with _____.

    many soldiers   many cities   no soldiers or weapons

2. **Confucianism** is the way of living based on the _____ of Confucius.

    language   teachings   alphabet

3. During a war, your **allies** will _____.

    be your enemies   attack your cities   help you fight

◆ Read and Remember

Write the Answer   Write one or more sentences to answer each question.

1. What kinds of landforms are on the Korean Peninsula?

2. When did Japan rule Korea?

3. How did the Korean War begin?

4. Where is the Demilitarized Zone?

5. What are North Korea's resources?

6. What are two problems in North Korea?

7. How is the economy of South Korea like that of Japan?

8. What kinds of products are made in South Korea?

9. How did the 1987 constitution help the people of South Korea?

10. What is the climate of Korea like?

11. What language do Koreans speak?

12. What two countries are Korea's northern neighbors?

13. What is the capital of North Korea?

14. What is the capital of South Korea?

15. In what year was Korea divided into North and South Korea?

16. What are the populations of North and South Korea?

◆ Think and Apply

Compare and Contrast   Copy the Venn diagram shown below on your paper. Then read each phrase below. Decide whether it tells about North Korea, South Korea, or both. Write the number of the phrase in the correct part of the Venn diagram on your paper. If the phrase tells about both nations, write the number in the center of the diagram on your paper.

1. a poor developing country

2. many natural resources

3. a rich industrial country

4. people speak the Korean language

5. land has many mountains

6. non-Communist government

7. farmland on the plains

8. Communist government

9. few natural resources

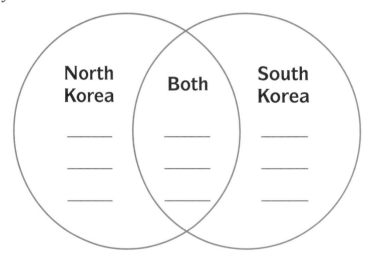

◆ Journal Writing

You have learned that North Korea and South Korea share the Korean Peninsula. They also shared the same culture for hundreds of years. Then Korea was divided and the countries fought a war that ended in 1953. The two countries are still divided today. However, South Korea has changed greatly. Write a paragraph in your journal that tells how life has improved in South Korea since 1953. Tell three or more ways the country has become a better place to live.

Reviewing a Physical Map

You learned that a **physical map** helps you find out about the elevation of a region. The physical map on this page shows where there are plains and mountains in Korea. The map shows the elevation of the mountains. Use the map key to know which colors show higher and lower elevations.

Study the map below. Then finish each sentence with an answer in dark print. Write the answer on your paper.

6,500–13,000	Seoul	southwest	10,000
9,000	lower	north	0–650

1. There are no mountains in Korea that are more than _____ feet.

2. Land near Mount Paektu has an elevation between _____ feet.

3. Mount Paektu has an elevation of about _____ feet.

4. From this map we can conclude that the elevation of Pyongyang is the same as the city of _____.

5. The elevation in the _____ is less than 650 feet.

6. The elevation of the city of Pusan is _____ than Kanggye.

7. The west coast of Korea has an elevation of _____ feet.

8. We can conclude that there are more mountains in the _____.

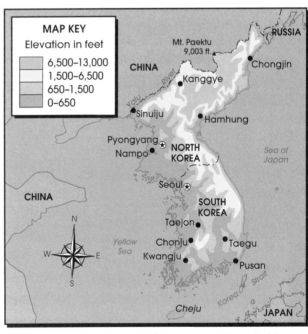

Physical Map of North Korea and South Korea

CHAPTER 45

Australia: A Country and a Continent

Where Can You Find?
Where can you find a huge rock that is more than one mile long?

Think About As You Read

1. What kind of region is Australia's outback?
2. Why did the British first settle in Australia?
3. Why are sheep important in both Australia and New Zealand?

New Words

- ◆ outback
- ◆ coral reef
- ◆ gold rush
- ◆ sparsely populated
- ◆ artesian wells
- ◆ stations
- ◆ two-way radios

People and Places

- ◆ Great Dividing Range
- ◆ Ayers Rock
- ◆ Great Barrier Reef
- ◆ Murray River
- ◆ Darling River
- ◆ Aborigines
- ◆ Perth
- ◆ Sydney
- ◆ Canberra
- ◆ Auckland

Which country is also a continent? Which country has more sheep than people? Which country is the home of many kinds of kangaroos? The answer is Australia.

Australia's Landforms

Australia is the driest and flattest continent. It is also the smallest continent. Australia is in the Southern Hemisphere. The seasons are the opposite of those in the Northern Hemisphere. When it is winter in the United States, it is summer in Australia.

The most important mountains in Australia are the Great Dividing Range. These low mountains begin in the northeast. They continue to the southeast. Fertile coastal plains are to the east and to the south of the mountains. These plains have most of Australia's cities and people. They have the best farmland.

The huge area to the west of the Great Dividing Range is covered with plains, plateaus, and low hills.

AUSTRALIA

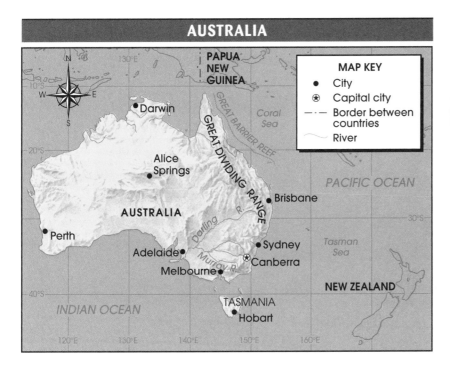

MAP KEY
- City
- ⊛ Capital city
- —·— Border between countries
- River

There are few large cities in Australia. What is the name of the only city on the west coast?

Australia's flag

Deserts cover at least one third of the country. The interior of Australia is called the **outback**.

One of the most interesting places in the interior is Ayers Rock. This huge red-brown rock is more than 1,000 feet high. It is more than one mile long.

The Great Barrier Reef is in the ocean near northeastern Australia. It is the world's largest **coral reef**. A coral reef is made from the skeletons of millions of tiny sea animals.

There are few rivers in Australia. The Murray and Darling rivers are the country's only long rivers.

Climate and Resources

Australia has many climates. Northern Australia is in the tropics. The north is always hot. Some places have tropical rain forests. The interior has a hot desert climate. The coastal plains near the Pacific Ocean get enough rain. All of Australia's large cities are on the coastal plains. Southern Australia has a temperate climate.

Australia is rich in natural resources. The country has coal, uranium, and bauxite. It exports more iron than any other nation. It has oil and natural gas. Australia exports minerals to Japan and other Asian countries.

Ayers Rock

The Great Barrier Reef is the largest coral reef in the world.

Australia has plants and animals that are not found in any other part of the world. Kangaroos and koalas are two of the animals that come from Australia.

History, Government, and Culture

The first people in Australia were the Aborigines. In 1788 the British started a colony in Australia for some of its prisoners. Later, British people who were not prisoners also settled in Australia. In 1851 gold was discovered. The country had a **gold rush**. People moved to the continent in order to find gold.

Australia became independent from Great Britain in 1901. Since then, it has been a member of the Commonwealth of Nations. Australia is a democracy. It is also a constitutional monarchy. Great Britain's queen, Queen Elizabeth, is also Australia's queen. A prime minister leads the country.

Australia is **sparsely populated**. It is a big country with only 18 million people. About one percent of the people are Aborigines. About four percent are Asian immigrants. Most Australians are white people. Their families once lived in Europe.

Koalas

Australian culture includes some of the British culture. English is the official language. Like the British, Australians drive their cars on the left side of the road. They celebrate Queen Elizabeth's birthday.

The Aborigines have their own culture. They have their own songs and music. They make beautiful paintings.

Cities and the Outback

Most Australians live in cities. There are five cities with more than one million people. Perth is the only city on the west coast. The largest cities are in the southeast. Sydney is the largest city. This southeastern port has almost 4 million people. Canberra, the capital, is also in this region.

Sydney harbor

The outback has underground water in **artesian wells**. The water in these wells comes to the surface without being pumped. This water is used to raise sheep and cattle.

People in the outback raise sheep and cattle on large ranches called **stations**. Australia has almost ten times more sheep than people. It exports more wool than any other nation.

Life can be lonely on a sheep station in the outback. A station can be hundreds of miles from the closest town. Most outback children live very far from all schools. So they study at home with a program called School of the Air. Special teachers use **two-way radios** to teach outback children who study at home.

The Economy

Australians enjoy a high standard of living. Australia is an industrial nation. It makes many kinds of products. Most of the factory products are used by Australians. They are not sold to other countries. Australia also earns a lot of money from tourism.

Working on a sheep station in the outback

Australia earns most of its money by exporting farm products. It sells wool, meat, and dairy products to many nations. Australia also exports wheat and fruit. Most of Australia's trade was once with Great Britain. Now most trade is with Japan, New Zealand, and the United States.

Many Aborigines have kept their old ways.

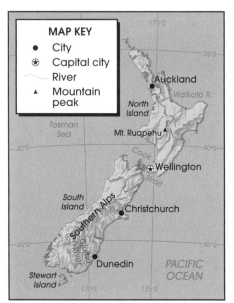

New Zealand

Australia is a flat, dry continent.

New Zealand: Australia's Nearest Neighbor

New Zealand is about 1,000 miles from Australia. The country has two large islands and some small ones. It is sparsely populated. Fewer than 4 million people live in New Zealand.

Like Australia, New Zealand was once a British colony. The British culture is important in New Zealand. Like Australia, the country raises millions of sheep and cattle. Most of its exports are farm products. New Zealand has a high standard of living.

Many parts of New Zealand are covered with mountains. The climate is cooler than Australia's. The country gets plenty of rain. Its largest city, Auckland, has only about 340,000 people.

Looking at Australia's Future

Australia is a large country with lots of dry, empty land. Perhaps one day Australia will find ways to bring water into the dry interior. Until that happens, Australia will be a large country with a small population.

Chapter Main Ideas

1. Australia is the flattest and driest continent. It has a lot of land but only 18 million people.
2. Most of Australia's interior has a desert climate. People raise sheep and cattle in the interior.
3. The Aborigines were the first Australians. Australia was a British colony until 1901.

◆ Vocabulary

Finish the Paragraph   Number your paper from 1 to 7. Use the word or words in dark print to finish the paragraph below. Write the word or words you choose on your paper.

sparsely populated	**two-way radios**	**coral reefs**	**artesian wells**
outback	**gold rush**	**stations**	

Many people moved to Australia after gold was found. A __1__ began in 1851. Australia's interior land that is away from the coast is called the __2__. The interior has few people. It is __3__. Many people raise sheep on large ranches called __4__. These ranches have __5__ where water comes to the surface without being pumped. Many children in the interior study at home and talk to the teachers of School of the Air by using __6__. Beautiful __7__ in the ocean are made from the skeletons of tiny sea animals.

◆ Read and Remember

Where Am I?   Read each sentence. Then look at the words in dark print for the name of the place for each sentence. Write the name of the correct place on your paper.

Great Dividing Range	**Auckland**	**Perth**	**New Zealand**
Murray and Darling	**Canberra**	**Sydney**	**Great Barrier Reef**

1. "I am in the capital of Australia."

2. "I am in the mountains in eastern Australia."

3. "I am at the two longest rivers in Australia."

4. "I am in Australia's largest city."

5. "I am in the only city on Australia's west coast."

6. "I am in New Zealand's largest city."

7. "I am in a country that is about 1,000 miles from Australia."

8. "I am visiting the world's largest coral reef."

◆ Think and Apply

Finding Relevant Information   Imagine you are telling your friend the reasons why Australia is sparsely populated. Read each sentence below. Decide which sentences are relevant to what you will say. On your paper write the relevant sentences you find. You should find four relevant sentences.

  1. Many places get very little rain.

  2. Deserts cover at least one third of Australia.

  3. Australia is an industrial nation.

  4. Sydney is a large port.

  5. Australia has plants and animals not found in any other part of the world.

  6. Australia has very few rivers.

  7. Ayers Rock is more than 1,000 feet tall.

  8. Australia is rich in natural resources.

  9. Australia has more sheep than people.

10. The interior has a hot desert climate.

◆ Journal Writing

You have been reading about the country and the continent of Australia. You have learned what the land, the climate, the cities, the animals, and the people are like. Imagine you are going to spend a week in Australia. Would you want to stay in a city or in the outback? Write a paragraph in your journal that tells where you would stay and why. Give three reasons for your choice.

Comparing Climate and Population Maps

We can learn more about Australia by comparing the **population** and **climate maps** on this page. The climate map shows a temperate, or mild, climate in the southeast. By comparing the two maps, we learn that more people live in the southeast where there is a temperate climate. Finish each sentence in Group A with an answer from Group B. Write the letter for the correct answer on your paper.

Group A

1. Most of Australia has a _____ climate.

2. Perth and Adelaide have a _____ climate.

3. Australia has _____ cities with a temperate climate.

4. Alice Springs has _____ people per square mile.

5. Some interior desert areas have _____.

6. _____ has a tropical climate.

7. The highest population density is in the temperate and _____ climates.

Group B

A. Darwin

B. 0–2

C. temperate

D. desert

E. no people

F. tropical

G. four

Population Map of Australia

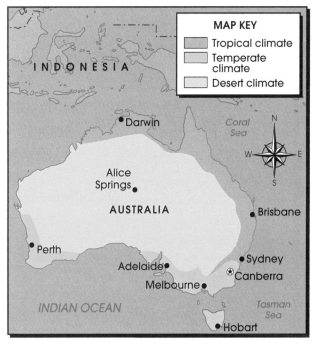

Climate Map of Australia

CHAPTER 46

The Pacific World: Oceania and Antarctica

Think About As You Read

1. What are the two kinds of islands in Oceania?
2. How do people earn a living in Oceania?
3. How is Antarctica different from other continents?

New Words

- uninhabited
- atolls
- lagoon
- glaciers
- icebergs
- penguins

People and Places

- Micronesia
- Guam
- Melanesia
- Papua New Guinea
- Fiji
- Polynesia

Imagine going on a long trip through the Pacific Ocean. You will travel through thousands of islands in Oceania. Then you will travel south for thousands of miles. Finally, you will reach the frozen continent of Antarctica.

Oceania: Three Groups of Islands

Oceania has three groups of islands. One group is Micronesia. Most of these islands are north of the Equator. Guam, an American territory, is an island in Micronesia.

The second group of islands is Melanesia. Most of these islands are south of the Equator. Papua New Guinea and Fiji are in Melanesia.

Polynesia is the largest group of islands. These islands are both north and south of the Equator. New Zealand and Hawaii are part of this group. Hawaii is a state in the United States.

The islands are different from each other in many ways. Some islands in Oceania are very small. Other

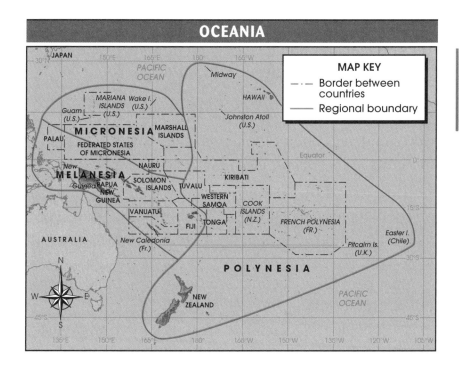

OCEANIA

MAP KEY
- –·– Border between countries
- —— Regional boundary

JAPAN

PACIFIC OCEAN

Midway

MARIANA ISLANDS Wake I. (U.S.)
Guam (U.S.)

HAWAII

MICRONESIA

MARSHALL ISLANDS

Johnston Atoll (U.S.)

PALAU

FEDERATED STATES OF MICRONESIA

Equator

NAURU

MELANESIA

New Guinea PAPUA NEW GUINEA

SOLOMON ISLANDS

TUVALU

KIRIBATI

WESTERN SAMOA

VANUATU

COOK ISLANDS (N.Z.)

TONGA

FIJI

FRENCH POLYNESIA (FR.)

New Caledonia (Fr.)

AUSTRALIA

Pitcairn Is. (U.K.)

Easter I. (Chile)

POLYNESIA

NEW ZEALAND

PACIFIC OCEAN

Oceania is divided into three regions, or groups of islands. What are the names of these regions?

islands are very large. New Zealand and New Guinea are the largest islands. The eastern part of New Guinea is called Papua New Guinea. You read in Chapter 39 that the western part of the island belongs to Indonesia.

About 13 million people live on thousands of islands in Oceania. Some islands have more than a million people. But most islands have fewer than 500,000 people. Many islands are **uninhabited**, or without people.

The Islands, Their Climates, and Their Resources

Oceania has two kinds of islands. There are high islands and low islands. High islands are made of mountains and volcanoes. Most high islands have fertile soil. Forests and jungles cover many high islands. Earthquakes and typhoons often cause great damage. Hawaii, New Zealand, and Guam are high islands.

Most of the low islands are called **atolls**. They are made of coral reefs. Often the coral reef surrounds a body of water. That water is called a **lagoon**. Atolls do not have fertile soil for farming. People who live on these islands get most of their food from the sea. Huge waves called tsunamis often cause great damage to the low islands.

Coral reef and atoll

Most Pacific islands are in the tropics. So they have a warm climate. Some islands receive very little rain. Other islands get plenty of rain for farming.

The islands of Oceania have few natural resources. Some of the islands get timber from their forests. New Guinea is one of the few islands that has some minerals. It has silver, gold, copper, and some oil.

Pineapples growing on a plantation in Hawaii

Culture and Education

The people of Oceania speak about 1,200 different languages. English is spoken more than any other language.

Most people in Oceania are Christians. Some people follow traditional island religions.

The people of Oceania enjoy their own traditional cultures. They have their own songs, dances, and clothing. But many people also enjoy western culture.

Most islands have elementary schools for their children. Some have high schools. Guam and Fiji are two of the islands that have universities. Thousands of students from other islands go to the university in Fiji.

Standard of Living and Earning a Living

The people of Hawaii and New Zealand enjoy high standards of living. These islands are the only developed areas in Oceania. Other islands have a low standard of living. On these developing islands, most people live in small villages.

Tourists enjoy the beach of Fiji.

There are very few ways to earn a living in most parts of Oceania. On the low islands, many people work at fishing. On the high islands, people work at both fishing and farming. They export such cash crops as sugarcane, pineapples, coffee, and coconuts.

Many people earn a living in Oceania through tourism. Thousands of people visit the beaches of Fiji, Guam, and other islands. Many islands could earn more money by developing their tourist industries.

Antarctica: The Coldest Continent

Antarctica is the continent at the South Pole. No one has a permanent home on this continent. A few thousand scientists work in Antarctica. They do research to learn about Earth.

Glaciers cover the tall mountains of Antarctica.

Antarctica is covered with thick ice. It has many tall mountains. Thick sheets of slow-moving ice called **glaciers** cover these mountains. There are also huge mountains of ice called **icebergs** in the nearby oceans.

Antarctica is really a frozen desert. In summer the temperature stays close to 32°F. In winter it can get as cold as 100°F below zero. The continent gets very little rain or snow.

Seals, whales, and **penguins** enjoy the icy ocean near Antarctica. Penguins are birds that cannot fly. They live on the land and swim in icy waters.

Antarctica has many minerals. But it is too difficult to mine minerals in this frozen land.

Oceania and Antarctica have few of the world's resources or factories. But they are important because they cover a large area of Earth.

Chapter Main Ideas

1. Micronesia, Melanesia, and Polynesia are three groups of islands in the Pacific.
2. Oceania has low islands made of coral reefs. It has high islands made of mountains and volcanoes.
3. Antarctica is a mountainous continent that is covered with thick ice.

Antarctica

Penguins

◆ Vocabulary

Finish Up   Choose the word in dark print that best completes each sentence. Write the word on your paper.

atolls   uninhabited   glaciers   lagoon   penguins   icebergs

1. Huge mountains of ice in the ocean are _____.

2. Low islands that are made of coral reefs are _____.

3. Many coral reefs surround a body of water called a _____.

4. A place that has no people is _____.

5. Birds that live in Antarctica and swim in its oceans are _____.

6. Huge slow-moving sheets of ice are _____.

◆ Read and Remember

Matching   Number your paper from 1 to 7. Each item in Group B tells about an item in Group A. Write the letter of each item in Group B next to the correct number on your paper.

Group A	Group B
1. Micronesia	**A.** These islands are made of mountains and volcanoes.
2. Polynesia	**B.** These workers do research in Antarctica.
3. scientists	**C.** This group of islands is north of the Equator.
4. Fiji and Guam	**D.** This type of island does not have fertile soil.
5. high islands	**E.** This is the largest island group in Oceania.
6. Hawaii	**F.** These islands have universities.
7. atoll	**G.** These Pacific islands are a state in the United States.

Write the Answer   Write one or more sentences to answer each question.

1. What island in Micronesia is an American territory?

2. About how many people live in Oceania?

3. How do people on high islands earn a living?

4. Why is Antarctica a desert?

5. What are the only two developed areas in Oceania?

6. How does tourism help the people of Oceania?

7. Why do scientists work in Antarctica?

◆ Think and Apply

Fact or Opinion   Number your paper from 1 to 10. Write **F** on your paper for each fact. Write **O** for each opinion. You should find five sentences that are opinions.

1. New Guinea and New Zealand are the largest islands in Oceania.

2. High islands have more fertile soil than low islands.

3. More people should settle on atolls.

4. Guam is a better place for tourists than Fiji.

5. Polynesia is the largest group of islands.

6. New Zealand and Hawaii are the most developed islands in Oceania.

7. The people of Papua New Guinea should work harder to mine their minerals.

8. Tourism is the best way for the Pacific islands to earn money.

9. People should start mining Antarctica's minerals.

10. There are no cities or villages in Antarctica.

◆ Journal Writing

Write a paragraph in your journal that tells why it is difficult to earn a living on most Pacific islands. Give at least two reasons.

CHAPTER 47

The Future of East Asia and the Pacific

Where Can You Find?
Where can you find a country whose people are working hard to protect the environment by recycling?

Think About As You Read

1. Where is overcrowding a problem in East Asia?
2. Why is a lack of democracy a problem in China and North Korea?
3. What are some nations doing to protect the environment?

New Words

- population growth rate
- human rights
- environment
- enforced

People and Places

- Tibet

Many countries in East Asia and the Pacific have strong economies. Some countries like Japan and Australia have very high standards of living. In some places people struggle to get enough food. This region must solve four problems so that all of its people can have a better life.

Overcrowded Nations

China, Japan, Hong Kong, Taiwan, and South Korea are all overcrowded. They do not have enough farmland. So it is difficult for some of the East Asian countries to feed their people. In Hong Kong and the cities of Japan most people live in small homes and apartments. In some cities in China, two families must sometimes share a tiny apartment.

East Asian countries are looking for ways to solve this problem. In Japan most people try to have small families. You learned that China's laws allow couples

A crowded apartment building in Shanghai, China

to have only one or two children. China's population is now growing more slowly. Its **population growth rate** has dropped to about one percent a year. But more than 15 million Chinese children are born each year. It will be very hard for China to feed such a large population.

A Lack of Democracy

China and North Korea are countries with dictators and Communist governments. Today China and North Korea do not allow **human rights**. Human rights are the rights that give people freedom and safety. People in China and North Korea cannot practice religion. They cannot speak or write against the government. In 1951 China took control of its southern neighbor, Tibet. It is now difficult for people in Tibet to practice their Buddhist religion. Now that Hong Kong is part of China, its people have less freedom to protest against the government.

The Chinese have destroyed many Buddhist shrines in Tibet.

China now trades with the United States, Japan, and other democratic nations. Perhaps these trading partners will pressure China to allow more freedom.

Communism ended in Eastern Europe and Russia in the early 1990s. Many people hope that communism will one day end in China and North Korea, too.

375

The Need to Protect the Environment

The nations of East Asia and the Pacific must find ways to protect their **environment**. The environment is the land, air, and water of a region.

Pollution has become a very big problem in China. China burns coal in order to make most of its energy. The country has air pollution from burning lots of coal. China also has acid rain. Pollution mixes with the rain and becomes acid rain. China's acid rain has damaged its forests, lakes, and rivers. Cities and villages in China also have water pollution problems. Much of the water in China is not safe to drink. China has passed laws to stop pollution. But it has not **enforced**, or carried out, these laws.

Japan also has pollution problems because it has so many cars and factories. Sometimes city air is so polluted that people wear special masks when they go outside. Japan has passed and enforced laws to protect the environment. Now the air and water are much less polluted.

The Japanese are also protecting their environment by recycling. They work hard to recycle as much as possible. They recycle paper, glass, plastics, and many other materials. They make new products from their old products. Recycling is helping Japan save important resources for the future.

Cars and factories cause pollution in the Pacific, too. Australia's big cities have pollution problems. Some of the cities on the Pacific islands are also starting to have pollution problems. In the years ahead, these countries must find ways to keep their air and water clean.

Keeping Old Traditions in Modern Nations

Traditions have always been important in East Asia. For hundreds of years, the teachings of Confucius guided the people of China and Korea. Many of these traditions have been lost under communism. People throughout Oceania have enjoyed different kinds of traditions. Can countries keep their traditions and still be modern nations? Japan has proved that it is possible.

Burning coal causes very bad air pollution in China.

The Japanese have been able to keep their ancient traditions and still be a modern nation.

These cans in Japan will be recycled to make new products.

Working for the Future

Most people in East Asia and the Pacific have a better life than their parents had. They have more food and better technology. They have better schools and industries.

You have read about many different countries and cultures. There are a lot of changes taking place in the world today. The Green Revolution is helping farmers in India grow enough food to feed its people. South Korea is becoming a more democratic country. Apartheid ended in South Africa in 1994.

Every day people in all parts of the world try to solve different kinds of problems. You have read about these problems in this book. Illiteracy, hunger, war, pollution, and a lack of freedom are some of the biggest problems. People in every country are working to have a better future. Perhaps you, too, will work for a better tomorrow for our world.

A good education for this schoolgirl in Fiji will give her a better future.

Chapter Main Ideas

1. Most countries in East Asia are overcrowded. It is difficult to feed their huge populations.
2. Millions of people in China and North Korea do not enjoy human rights.
3. As nations have more cars and factories, they have more pollution problems.

◆ Vocabulary

Match Up   Finish the sentences in Group A with words from Group B. Write the letter of each correct answer on your paper.

Group A

1. The _____ tells you how much a population grows and changes during a year.

2. The rights that protect the freedom and safety of people are called _____.

3. The land, air, and water of a region are the _____.

4. A law is _____ when it is carried out.

Group B

A. human rights

B. population growth rate

C. enforced

D. environment

◆ Read and Remember

Complete the Chart   Copy the chart shown below on your paper. Use the facts from the chapter to complete the chart. You can read the chapter again to find facts you do not remember.

Problems of East Asia and the Pacific

	What Is the Problem?	How Does the Problem Hurt the Region?	What Is One Way the Problem Is Being Solved?
1.			
2.			
3.			

Write the Answer   Write one or more sentences to answer each question.

1. How many children are born in China each year?

2. What country was taken over by China in 1951?

3. What are some ways people in East Asia and the Pacific have better lives than their parents had?

◆ Think and Apply

Find the Main Idea   On your paper copy the boxes shown below. Then read the five sentences below. Choose the main idea and write it on your paper in the main idea box. Then find three sentences that support the main idea. Write them on your paper in the boxes of the main idea chart. There will be one sentence in the group that you will not use.

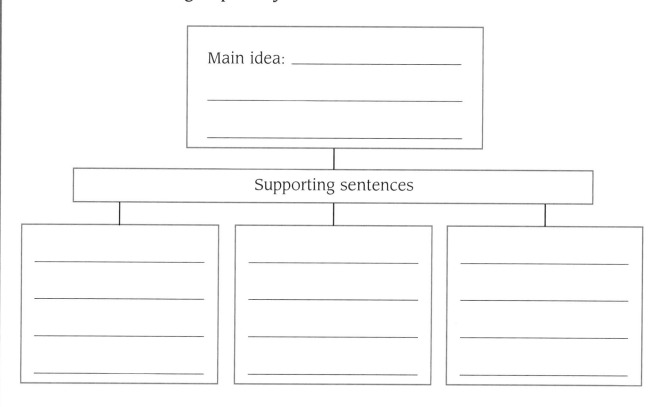

a. The Chinese government does not allow Buddhists in Tibet to practice their religion.

b. China has serious pollution problems.

c. China does not allow people to speak or write against the government.

d. China does not allow democracy and human rights.

e. The Chinese government stopped the student protests at Tiananmen Square.

◆ Journal Writing

Write a paragraph about one problem in East Asia and the Pacific. Then tell two or three ways nations may solve that problem.

GREENLAND
(DEN.)

RUSSIA

ALASKA
(U.S.)

CANADA

PACIFIC
OCEAN

UNITED STATES

ATLANTIC
OCEAN

MIDWAY
ISLANDS
(U.S.)

WAKE
ISLAND
(U.S.)

HAWAII
(U.S.)

MEXICO

Inset, below left

VENEZUELA
GUYANA
SURINAME
FRENCH GUIANA
(FR.)

CAPE
VERDE

GUATEMALA
EL SALVADOR

MARSHALL
ISLANDS

TOKELAU
(N.Z.)

WESTERN
SAMOA

COLOMBIA

ECUADOR

Equator

GALÁPAGOS IS.
(ECUA.)

NAURU

KIRIBATI

BRAZIL

SOLOMON
ISLANDS

TUVALU

PERU

VANUATU

COOK
IS.
(N.Z.)

FRENCH
POLYNESIA
(FR.)

BOLIVIA

PARAGUAY

NEW
CALEDONIA
(FR.)

TONGA

FIJI

NIUE
(N.Z.)

AMERICAN
SAMOA (U.S.)

PACIFIC
OCEAN

CHILE

URUGUAY

ARGENTINA

NEW
ZEALAND

FALKLAND IS.
(U.K.)

SOUTH GEORGIA
(U.K.)

Caribbean Inset

Gulf of Mexico

U.S.

BAHAMAS

ATLANTIC
OCEAN

TURKS &
CAICOS ISLANDS
(U.K.)

MEXICO

CUBA

CAYMAN IS.
(U.K.)

PUERTO
RICO
(U.S.)

VIRGIN ISLANDS
(U.K.)

DOMINICAN
REPUBLIC

HAITI

BELIZE

JAMAICA

ANTIGUA & BARBUDA

GUATEMALA

VIRGIN IS. (U.S.)

GUADELOUPE (FR.)

HONDURAS

ST. KITTS
& NEVIS

DOMINICA

MARTINIQUE (FR.)

CURAÇAO
(NETH.)

ANTARCTICA

EL
SALVADOR

NICARAGUA

ARUBA
(NETH.)

ST. LUCIA

BARBADOS

GRENADA

ST. VINCENT &
THE GRENADINES

Caribbean Sea

COSTA RICA

Panama
Canal

PANAMA

BONAIRE
(NETH.)

TRINIDAD &
TOBAGO

PACIFIC
OCEAN

COLOMBIA

VENEZUELA

GUYANA

Inset, below right

RUSSIA

KAZAKSTAN

MONGOLIA

UZBEKISTAN

KYRGYZSTAN

TURKMENISTAN

TAJIKISTAN

NORTH KOREA

JAPAN

CHINA

Mediterranean Sea

IRAQ

IRAN

AFGHANISTAN

BHUTAN

SOUTH KOREA

PACIFIC OCEAN

OROCCO

KUWAIT

BAHRAIN

PAKISTAN

NEPAL

HONG KONG

NORTHERN MARIANA ISLANDS (U.S.)

ALGERIA

LIBYA

EGYPT

QATAR

SAUDI ARABIA

UNITED ARAB EMIRATES

INDIA

MYANMAR

LAOS

TAIWAN

SAHARA (MOR.)

OMAN

MACAO

AURITANIA

MALI

NIGER

CHAD

ERITREA

YEMEN

BANGLADESH

THAILAND

VIETNAM

PHILIPPINES

GUAM (U.S.)

8

SUDAN

DJIBOUTI

CAMBODIA

BRUNEI

4

NIGERIA

11

ETHIOPIA

SRI LANKA

MALDIVES

PALAU

FEDERATED STATES OF MICRONESIA

5

7

9

15

6

10

12

SOMALIA

MALAYSIA

EQUATORIAL GUINEA

UGANDA

KENYA

SINGAPORE

PAPUA NEW GUINEA

SÃO TOMÉ & PRÍNCIPE

13

14

RWANDA

BURUNDI

16

INDONESIA

SOLOMON ISLANDS

CABINDA (ANG.)

TANZANIA

SEYCHELLES

ATLANTIC OCEAN

MALAWI

ANGOLA

COMOROS

ZAMBIA

ZIMBABWE

MADAGASCAR

INDIAN OCEAN

1 SENEGAL
2 GAMBIA
3 GUINEA-BISSAU
4 GUINEA
5 SIERRA LEONE
6 LIBERIA
7 CÔTE D'IVOIRE
8 BURKINA FASO
9 GHANA
10 TOGO
11 BENIN
12 CAMEROON
13 GABON
14 REPUBLIC OF CONGO
15 CENTRAL AFRICAN REPUBLIC
16 DEMOCRATIC REPUBLIC OF CONGO

NAMIBIA

BOTSWANA

MAURITIUS

MOZAMBIQUE

AUSTRALIA

SOUTH AFRICA

SWAZILAND

LESOTHO

N

W

E

S

Europe Inset

NORWAY

SWEDEN

ESTONIA

NETHERLANDS

(RUSSIA)

LATVIA

RUSSIA

DENMARK

LITHUANIA

IRELAND

UNITED KINGDOM

GERMANY

POLAND

BELARUS

BELGIUM

CZECH REPUBLIC

SLOVAK REPUBLIC

UKRAINE

LUXEMBOURG

LIECH.

ATLANTIC OCEAN

SWITZ.

AUSTRIA

HUNGARY

MOLDOVA

FRANCE

SAN MARINO

SLOVENIA

CROATIA

ROMANIA

BOSNIA AND HERZ.

YUGO-SLAVIA

BULGARIA

ANDORRA

MONACO

ITALY

MACEDONIA

GEORGIA

PORTUGAL

SPAIN

VATICAN CITY

ALBANIA

GREECE

ARMENIA

TURKEY

AZERBAIJAN

GIBRALTAR (U.K.)

MALTA

CYPRUS

SYRIA

IRAN

TUNISIA

Mediterranean Sea

LEBANON

IRAQ

MOROCCO

ALGERIA

ISRAEL

(WEST BANK)

LIBYA

EGYPT

JORDAN

381

ARCTIC OCEAN

PACIFIC OCEAN

ASIA

URAL MOUNTAINS

Huang He

Chang Jiang

HIMALAYAS

Indus River

Ganges River

EUROPE

ALPS

Danube River

Mediterranean Sea

Niger River

SAHARA DESERT

AFRICA

Nile River

Congo River

Arabian Sea

INDIAN OCEAN

AUSTRALIA

Darling River

ATLANTIC OCEAN

NORTH AMERICA

ROCKY MOUNTAINS

Mississippi River

APPALACHIAN MOUNTAINS

Gulf of Mexico

Caribbean Sea

Amazon River

SOUTH AMERICA

ANDES

ATLANTIC OCEAN

PACIFIC OCEAN

Bering Sea

Equator

N
E
S
W

ANTARCTICA

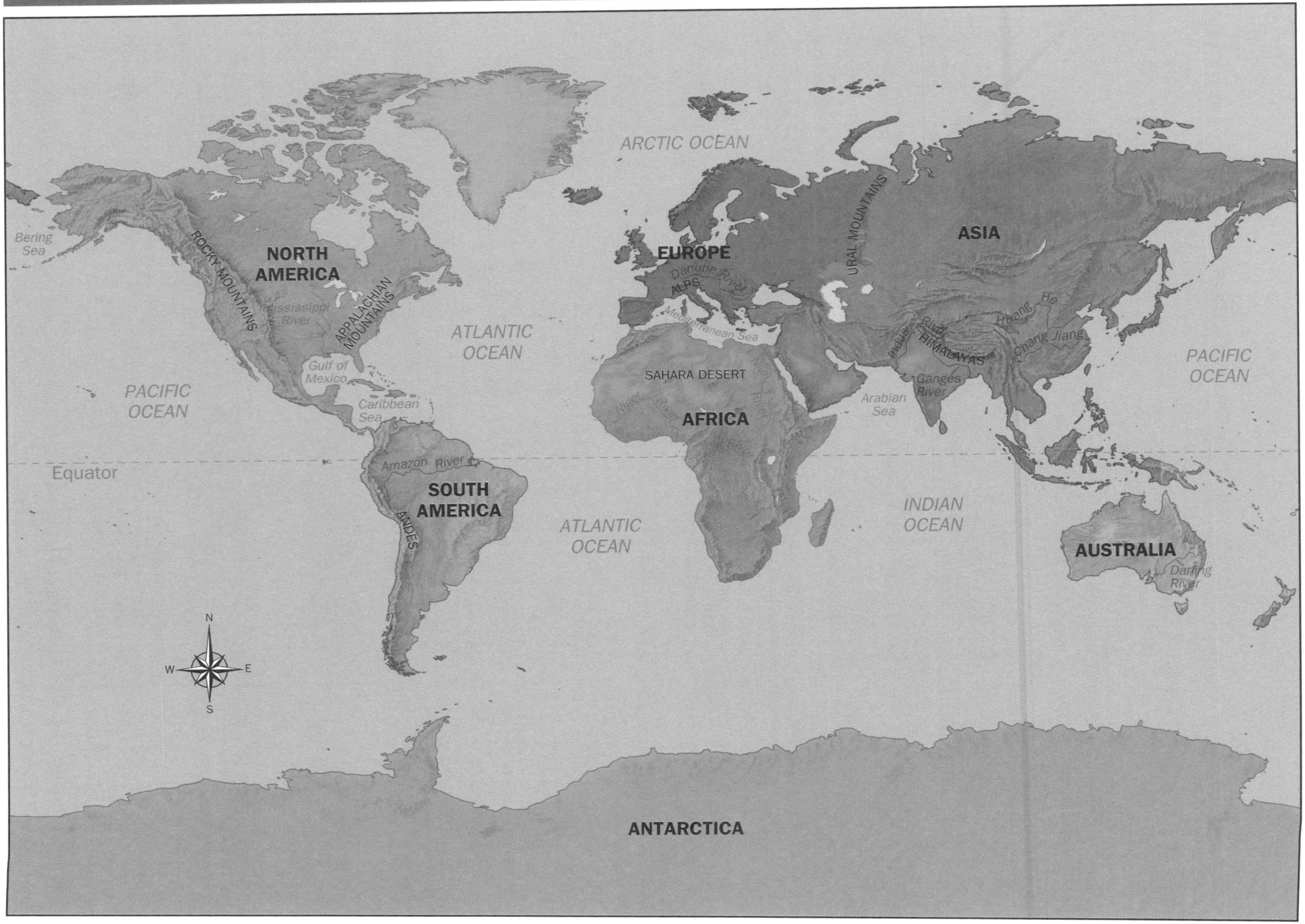

aba (page 276)   An aba is a long black robe that covers a Muslim woman's head, face, and body.

absolute monarchy (page 275)   In an absolute monarchy, the king or queen has full power to make all laws.

acid rain (page 35)   Acid rain forms when pollution in the air becomes part of the rain or snow.

Act of Union (page 106)   The Act of Union joined England, Wales, and Scotland into one nation in 1707.

acupuncture (page 339)   Acupuncture is a system that uses small needles put into different parts of the body to treat disease and pain.

agriculture (page 100)   Agriculture is another name for farming, growing crops, and raising livestock.

AIDS (page 234)   AIDS is a serious disease that has no cure.

alcohol abuse (page 163)   Alcohol abuse means people drink too many alcoholic drinks.

ally (page 355)   An ally is a country that joins another country in a common cause.

aluminum (page 66)   Aluminum is a lightweight, shiny metal made from bauxite.

ambassador (page 83)   An ambassador is a representative of his or her government in another country.

ancient (page 251)   Ancient means very old.

apartheid (page 225)   Apartheid laws in South Africa kept racial groups apart. Apartheid ended in 1991.

aquifer (page 282)   An aquifer is an area of underground water.

arch (page 132)   An arch is a curved opening in doors, bridges, or tunnels.

archipelago (page 312)   An archipelago is a chain of islands.

artesian well (page 363)   Water in an artesian well comes to the surface without being pumped.

assemble (page 203)   To assemble means to put parts together.

atoll (page 369)   An atoll is a low island made of coral reefs.

average (page 307)   Average means usual.

Bahasa Indonesia (page 315)   Bahasa Indonesia is the official language of Indonesia.

basin (page 74)   A basin is the area of land that drains into a river.

bauxite (page 66)   Bauxite is an ore used to make aluminum.

Berlin Wall (page 122)   The Berlin Wall separated East Berlin and West Berlin.

billionaire (page 219)   A billionaire is a person who has more than one billion (one thousand million) dollars.

birth defect (page 187)   A birth defect is a problem of the mind or body with which some babies are born.

boat people (page 307)   Boat people were non-Communist refugees who escaped from Vietnam in small boats at the end of the Vietnam War.

bonsai (page 346)   A bonsai is a tiny tree grown in a pot, usually in Japanese gardens.

British Commonwealth (page 28)   The British Commonwealth is a group of nations once ruled by Great Britain.

Buddhism (page 290)   Buddhism is a religion based on the teachings of Buddha.

bullet train (page 114)   A bullet train travels faster than 180 miles an hour.

café (page 115)   A café is a sidewalk restaurant popular in France.

calypso (page 66)   Calypso is a type of music that mixes African, Spanish, and American cultures in the Caribbean area.

canal (page 36)   A canal is a human-made waterway dug across land to join two bodies of water.

cape (page 223)   A cape is a point of land that sticks out into a large body of water.

Carnaval (page 73)   Carnaval is a holiday in Brazil that is celebrated with costumes and parades.

cash crop (page 17)   A cash crop is a crop that can be sold, especially a crop that is exported.

cassava (page 202)   A cassava is an African vegetable.

caste (page 297)   A caste is a group of people in the Hindu religion.

caste system (page 297)   The caste system is the way people in India are divided into groups.

cease-fire (page 180)   A cease-fire is an agreement to stop fighting.

cemetery (page 281)   A cemetery is a place where people are buried.

central government (page 180)   The central government is the main government in a country that has other parts with governments, such as states.

chancellor (page 122)   Chancellor is the title of the leader of Germany.

character (page 339)   A character is a little picture that stands for a word or an idea in Japanese, Chinese, and other languages.

chemical fertilizer (page 250)   A chemical fertilizer is a mixture of chemicals added to soil to make plants grow better.

chernozem (page 173)   Chernozem is rich, fertile, black soil in the Ukraine.

Christianity (page 243)   Christianity is the Christian religion.

chromite (page 267)   Chromite is a mineral used to make chrome, a shiny metal that does not rust.

citizen (page 15)   A citizen is a member of a country.

civil war (page 58)   During a civil war, people of the same country fight against each other.

climate (page 10)   Climate is the kind of weather a place usually has over a period of time.

coastal plain (page 9)   A coastal plain is flat land near an ocean.

coca (page 82)   Coca is a plant grown to make cocaine.

cocaine (page 82)   Cocaine is an illegal drug made from the coca plant.

collective (page 337)   A collective is a huge government-owned farm in China.

colony (page 48)   A colony is land that is ruled by another nation.

colored (page 225)   A colored person in South Africa has black and white or black and Asian parents.

command economy (page 146)
In a command economy, the government controls everything and decides what factory products to make, what prices to charge, what wages to pay, and what crops to grow.

commercial agriculture (page 292)
Commercial agriculture is a type of farming in which cash crops are grown on large plantations.

Commonwealth of Independent States (CIS) (page 144)   The CIS is an organization of Russia and 11 other former Soviet Union republics.

communicate (page 315)   To communicate means to give information.

communism (page 92)   Under communism the government owns and controls all of a country's farms, factories, businesses, and money.

Communist (page 58)   Under a Communist system, the government controls everything in a nation.

Communist party (page 307)
The Communist party is the government of a Communist country.

concentration camp (page 153)
Concentration camps were death camps built by the Germans during World War II to kill Jews and other minorities.

conflict (page 219)   A conflict is a fight that often lasts a long time.

Confucianism (page 354)   Confucianism is a way of living based on the teachings of Confucius.

conservation (page 282)   Conservation means to save natural resources.

constitutional monarchy (page 100)
A constitutional monarchy is a democracy that has a king or queen with little power.

consumer good (page 153)
A consumer good is something people buy to eat or to use.

contaminated (page 172)
Contaminated means not pure or polluted with something undesirable.

continent (page 4)   A continent is a large body of land on Earth.

continental climate (page 148)   In a continental climate, winters are long and cold. Summers are short and hot.

contract (page 337)   A contract is a legal agreement between two groups that tells what each group can do.

coral reef (page 361)   A coral reef is made from the skeletons of millions of tiny sea animals.

criticize (page 184)   To criticize is to complain about or to find fault with something.

crossroads (page 240)   A crossroads is the place where roads meet.

culture (page 1)   Culture is the beliefs, ideas, art, and customs of a group of people.

currency (page 123)   Currency is the money used by a country.

czar (page 145)   Czars were the kings of Russia before the Russian Revolution.

defeat (page 121)   To defeat means to win a fight.

deforestation (page 75)   Deforestation is the act of destroying a forest by cutting down and burning trees.

delta (page 201)   A delta is land made of soil that a river has carried and left at the place where the river flows into the sea.

Demilitarized Zone (page 355)
The DMZ is a strip of land that separates North Korea from South Korea at the 38th parallel.

democracy (page 10)   In a democracy the people vote for their leaders in free elections.

democratic (page 60)   Democratic means a way of doing things that gives people freedom.

demonstration (page 336)
A demonstration is a public meeting where groups of people show how they feel.

deposited (page 249)   Soil is deposited by a river when it is left behind.

desalination plant (page 277)
A desalination plant takes the salt out of ocean water to make fresh water.

desertification (page 232)
Desertification occurs when deserts grow larger and grasslands become smaller.

developed nation (page 12)
A developed nation is an industrial nation with a high standard of living.

developing nation (page 43) A developing nation has a low standard of living and poor industry or technology.

dictator (page 58) A dictator has full power to make laws and control a country's land and money.

dike (page 338) A dike is a special wall built to control floods.

disease (page 234) Disease means sickness.

dormant (page 349) Dormant means quiet. A dormant volcano may erupt again.

drip irrigation (page 257) Drip irrigation conserves water by using hoses to water individual plants.

drought (page 194) A drought is a long period of time without rain.

drug trafficker (page 82) A drug trafficker ships illegal drugs into countries where the illegal drugs are sold.

earthquake (page 56) During an earthquake the ground shakes and often cracks open.

Eastern Hemisphere (page 67) The Eastern Hemisphere is the half of Earth with Europe, Asia, Africa, and Australia.

elevation (page 42) Elevation tells how high land is above sea level.

empire (page 81) An empire is a group of nations that is ruled by one country under one leader.

enforced (page 376) Enforced means carried out.

environment (page 376) The environment is the land, air, and water of a region.

Equator (page 41) The Equator is the 0° line of latitude that divides the globe into north and south.

erosion (page 233) Erosion takes place when soil is blown away by wind or washed away by rain.

escarpment (page 224) An escarpment is a steep cliff.

ethnic cleansing (page 179) Ethnic cleansing occurred in Bosnia when the people of one ethnic group removed and killed people from other ethnic groups.

ethnic group (page 106) An ethnic group is a group of people of the same culture.

euro (page 140) The euro will be the currency of the European Union in 1999.

European Community (page 140) The European Community was formed by countries of Western Europe to work together to improve trade.

European Union (page 140) The European Union, or EU, is a trade group of 15 European countries.

expert (page 234) An expert is a person with special skills or knowledge.

export (page 33) To export is to sell goods to other countries.

extended family (page 297)   In an extended family, relatives such as grandparents, aunts, uncles, cousins, children, and parents all live together.

famine (page 232)   A famine is a terrible shortage of food for a long period of time.

favorable balance of trade (page 132)   A favorable balance of trade means a country exports more than it imports.

ferry boat (page 313)   A ferry boat carries people across small bodies of water.

fertile (page 98)   Soil that is fertile is good for farming.

fertilizer (page 100)   A fertilizer is a mixture of materials that makes soil better for growing plants.

foreign (page 275)   People are foreign if they come from another country.

foreign aid (page 234)   Foreign aid is money and help that one nation gives to another nation.

foreign investment (page 322)   A foreign investment is made when business owners use their money to start businesses in other countries.

free enterprise (page 331)   Free enterprise means people can own and control their own businesses.

free market economy (page 147)   In a free market economy, citizens and companies own and control factories, farms, and businesses.

freedom of religion (page 10)   Freedom of religion means people can pray any way and any place they choose.

geography (page 1)   Geography is the study of Earth's people, landforms, climates, and resources.

geothermal energy (page 139)   Geothermal energy uses steam from hot places inside the earth to do work.

glacier (page 371)   A glacier is a thick sheet of slow-moving ice that covers the mountains of Antarctica.

glasnost (page 162)   Glasnost is the Russian word for the idea of openness and freedom begun by Mikhail Gorbachev.

gold rush (page 362)   In a gold rush, people move to an area where gold has been discovered.

gorge (page 236)   A gorge is a deep valley with steep rocky walls.

grassland (page 194)   A grassland is an area covered with grass.

Green Revolution (page 322)   The Green Revolution means farmers in developing countries use miracle seeds and modern farming methods to grow large crops.

Hajj (page 276)   A Hajj is the visit to Mecca that all Muslims must make at least once.

heavy industries (page 162)   Heavy industries make machines, steel, weapons, and other large products.

Hebrew (page 243)   Hebrew is the official language of Israel.

herb (page 339)   An herb is a plant whose parts are used to make medicines.

highlands (page 56)   Highlands are hills and low mountains.

Hindi (page 296)   Hindi is the main official language of India.

Hinduism (page 290)   Hinduism is the main religion of India.

Holocaust (page 124)   The Holocaust was the German killing of Jews and other people in concentration camps during World War II.

homeland (page 259)   A homeland is a person's native country.

homeless (page 76)   A person who is homeless does not have a place to live.

human rights (page 375)   Human rights are the rights that give people freedom and safety.

human/environment interaction (page 2)   Human/environment interaction is the geography theme that tells how people use, change, and work with a place.

humid (page 314)   A humid climate is damp.

hunger (page 231)   Hunger means there is not enough food.

hurricane (page 65)   A hurricane is a tropical storm with very strong winds and heavy rain.

hydroelectric power (page 139)   Hydroelectric power is electricity made using the power of falling water.

iceberg (page 371)   An iceberg is a huge mountain of ice in the ocean.

illegal drug (page 51)   An illegal drug is a substance of which the use and sale are against the law.

illegal immigrant (page 50)   An illegal immigrant is a person who moves into a country without permission.

illiteracy (page 90)   Illiteracy means people cannot read and write.

immigrant (page 10)   An immigrant is a person who moves into a country from another country.

import (page 33)   To import is to buy goods from other countries.

independent (page 66)   Independent means people rule themselves.

industrial nation (page 11)   An industrial nation has factories that use modern technology to make things.

Industrial Revolution (page 106)   The Industrial Revolution changed industry from making things by hand at home to making them using machines in factories.

inflation (page 58)   Inflation means goods are more expensive to buy.

inland (page 330)   An inland area is far from the ocean.

interior (page 74)   The interior is the inside, often undeveloped, area of a country away from the borders or coast.

investing (page 308)   Investing is when business owners use their money to start other businesses and develop an area.

irrigate (page 187)   To irrigate means to bring water to fields.

Islam (page 178)   Islam is the world's second largest religion, practiced mostly in Asia, Africa, and the Middle East.

Islamic fundamentalism (page 248)   Under Islamic fundamentalism, Muslims follow the strictest rules of Islam.

isthmus (page 59)   An isthmus is a narrow piece of land that is between two large bodies of water.

Judaism (page 242)   Judaism is the religion of the Jews.

jute (page 299)   Jute is used to make rope.

Kaaba (page 276)   The Kaaba is the holiest place in Islam. It is located in the Great Mosque at Mecca.

kibbutz (page 258)   A kibbutz is a farm in Israel owned by its members who share all of the work.

kimono (page 346)   A kimono is a traditional Japanese robe.

Koran (page 243)   The Koran is the holy book of Muslims. Muhammad's teachings are in the Koran.

lagoon (page 369)   A lagoon is a body of water surrounded by a coral reef.

land reform (page 89)   Land reform occurs when a government divides large plantations into smaller farms that are given to poor farmers.

landform (page 3)   A landform is the shape of an area on Earth's surface. Plains, plateaus, hills, and mountains are the four main landforms.

Latin (page 40)   Latin is a language that few people speak today. The Spanish and Portuguese languages came from Latin.

latitude (page 2)   Lines of latitude are imaginary lines that run east to west and measure distance in degrees north or south of the Equator.

lava (page 56)   Lava is melted rock that pours out from a volcano.

legal (page 154)   Something that is legal is allowed by law.

Lingala (page 218)   Lingala is an African language spoken in Congo.

loan (page 234)   A loan is money that one nation lends another.

location (page 2)   Location is the geography theme that tells where a place is located, such as near another place. Lines of latitude and longitude give the exact location of a place.

logging (page 26)   Logging is cutting down trees in a forest.

longitude (page 2)   Lines of longitude are imaginary lines that meet at the North and South Pole and measure distance in degrees east or west of the Prime Meridian.

lower class (page 50)   The lower class is a group of poor people who often live in small, poorly made homes in slums or rural villages.

lowlands (page 26)   Lowlands are low, flat land usually near lakes or oceans.

mainland (page 329)   A mainland is the main area of a region, not including the islands off its coast.

majority (page 179)   A majority means more than half.

malnutrition (page 219)   Malnutrition is poor health caused by lack of good food.

manganese (page 252)   Manganese is a metal ore.

manufacture (page 114) To manufacture means to make something from raw materials usually in a factory using machinery.

manufactured goods (page 33) Manufactured goods are goods that are made in factories from other materials.

manufacturing (page 307) Manufacturing means making products in a factory.

marijuana (page 84)   Marijuana is a plant that is used as an illegal drug.

Mediterranean climate (page 112) The Mediterranean climate has long, hot, dry summers and short, rainy winters.

mestizo (page 42)   A mestizo is a person in Latin America who has European and Indian ancestors.

middle class (page 50)   The middle class is a group of people who are not rich and not poor.

minority (page 225)   A minority is a small part of the population.

miracle seeds (page 322)   Miracle seeds are special seeds that produce larger crops.

missile (page 283)   A missile is a weapon with bombs.

modest (page 276)   To be modest means to be proper.

monarch (page 106)   A monarch is a king or a queen.

monotheism (page 242)   Monotheism is the belief in one God.

monsoon (page 288)   A monsoon is a seasonal wind.

monument (page 133)   A monument is a structure built to remember an important person or event.

mosque (page 171)   A mosque is a place of the Islam religion where Muslims pray.

mountain chain (page 9)   A mountain chain is a long group of mountains, the landform with high elevation, often with steep, rocky sides.

mountainous (page 130)   Mountainous means mountains cover a large part of a region.

mouth (page 72)   The mouth of a river is where the river flows into an ocean.

movement (page 2)   Movement is the geography theme that tells how and why people, ideas, and goods move from place to place.

mulatto (page 73)   A mulatto is a person with black African and white European ancestors.

NAFTA (page 34)   NAFTA is the North American Free Trade Agreement between the United States, Canada, and Mexico to improve trade.

national anthem (page 121)   A national anthem is a special song that expresses love or pride for one's country.

national debt (page 50)   A country's national debt is the money that it has borrowed and must repay.

national park (page 211)   A national park is a government-owned area where wild animals are protected.

nationalism (page 179)   Nationalism means a very strong love or pride for a person's country or ethnic group.

NATO (page 137)   NATO is the North Atlantic Treaty Organization formed by 13 Western European nations, Canada, Turkey, and the United States to protect these countries from communism.

natural disaster (page 330)   A natural disaster is a terrible event caused by nature. Volcanoes, earthquakes, typhoons, and tsunamis are all natural disasters.

natural gas (page 105)   Natural gas is a natural resource that is used as fuel.

natural resource (page 16)   A natural resource is something in nature that people use such as coal, oil, natural gas, minerals, fish, farmland, and forests.

newsprint (page 26)   Newsprint is paper made from wood and used for making newspapers.

nomad (page 209)   A nomad moves from place to place looking for food and water for his animals.

Northern Hemisphere (page 41) The Northern Hemisphere is the half of Earth north of the Equator.

nuclear energy (page 114)   Nuclear energy is derived from uranium and is used to generate electricity.

nuclear power plant (page 114) A nuclear power plant is used to generate electricity from nuclear energy.

nuclear weapon (page 170)   A nuclear weapon is a powerful bomb that can destroy a large city.

oasis (page 242)   An oasis is a place in the desert that has underground water.

ocean current (page 98)   An ocean current is a strong movement of water in the ocean.

official language (page 8)   An official language is the legal language used for all the business of a country.

oil refinery (page 211)   In an oil refinery, oil is cleaned and changed into products such as gasoline.

okapi (page 217)   The okapi is a furry animal that lives only in Congo's rain forests. It is the symbol of Congo.

one-crop economy (page 91)   In a one-crop economy, a country earns most of its money from one crop.

OPEC (page 276)   OPEC, the Organization of Petroleum Exporting Countries, is an organization of countries that export oil.

organization (page 169) An organization is a group that joins together for a specific purpose.

orphan (page 300)   An orphan is a child whose parents are dead.

orphanage (page 300)   An orphanage is a group home for orphans.

Ottoman Empire (page 266) The Ottoman Empire was the area of the Middle East and North Africa controlled by the Ottoman Turks from the 1500s until after World War I.

outback (page 361)   The outback is the interior of Australia.

overcrowded (page 204)   A place is overcrowded when it has too many people.

overgraze (page 233)   Overgrazing is when sheep and cattle destroy an area by eating all the grasses.

overpopulation (page 281) Overpopulation means there are too many people living in a region.

oxygen (page 75)   Oxygen is a gas released by plants that most animals need to breathe.

Palestinian Liberation Organization (PLO) (page 260)   The PLO is an organization that wants a homeland for the Palestinians in Israel.

parallel (page 306)   A parallel is a line of latitude.

Parliament (page 28)   Parliament is the name of a group of lawmakers in certain democracies.

Partnership for Peace (page 138) The Partnership for Peace, or PfP, is formal cooperation between NATO and former Communist countries in Eastern Europe.

passport (page 137)   A passport is a document given by a government to citizens of that country so they can travel to other countries.

Peace Corps (page 235)   The Peace Corps is an American group that sends its members to help developing countries.

penguin (page 371)   A penguin is a bird that cannot fly. Many penguins enjoy the icy ocean near Antarctica.

peninsula (page 96)   A peninsula is land that has water on three sides.

permafrost (page 25)   Permafrost is the layer of Arctic and subarctic soil that is always frozen.

petroleum (page 105)   Petroleum is another name for oil.

pharaoh (page 251)   The pharaohs were the kings of ancient Egypt.

phosphate (page 244)   Phosphate is a natural resource used to make fertilizers.

place (page 2)   Place is the geography theme that tells how an area is different from other areas.

plantation (page 43)   A plantation is a huge farm, especially in the tropics, that grows cash crops such as bananas, coffee, and sugarcane.

plateau (page 3)   A plateau is a broad area of high, flat land. It is one of the four basic landforms.

plot (page 162)   A plot is an area of land, usually small, used for farming.

political unrest (page 186)   During political unrest, people are unhappy with their government and want to change it.

pollution (page 35)   Pollution is waste materials, such as smoke, dirt, trash, and chemicals, that make air and water impure, or not clean.

pope (page 133)   The pope is the world leader of the Roman Catholic Church and the ruler of Vatican City.

population density (page 16)   Population density tells how crowded a place is based on the number of people living in a certain size region.

population growth rate (page 375)   The population growth rate is the number of children born each year as a percent of the total population.

poverty (page 76)   Poverty is the condition of being very poor with a very low standard of living.

prime minister (page 28)   The prime minister is one type of leader of the government in a democracy.

privatization (page 154)   Under privatization a government sells farms and factories it owns to citizens.

profit (page 337)   Profit is the money made after subtracting expenses.

protest (page 336)   To protest is to complain strongly against something.

province (page 24)   A province is a part of a nation, similar to a state, that has its own government.

pyramid (page 251)   The pyramids are large stone structures built in ancient Egypt.

Quechua (page 81)   Quechua is an Indian language that is one of the two official languages of Peru.

racial group (page 225)   A racial group is a group of people who share physical characteristics such as skin color.

radioactive waste (page 172) Radioactive waste is contaminated materials left after making nuclear energy.

rapid population growth (page 89) Rapid population growth means the population grows a lot each year.

raw material (page 97) Raw material is natural resources used to make manufactured products.

rebel (page 220) To rebel means to fight against the government.

recycle (page 139) To recycle means to use again for the same or a different purpose.

reelected (page 315) To be reelected means to be voted into office again.

refugee (page 219) A refugee is a person who leaves home during a war.

region (page 2) Region is the geography theme that describes areas based on the area's climate, landforms, culture, or something else in common.

religion (page 1) Religion is the way people believe and pray to God or to many gods.

religious order (page 300) A religious order is a group of people who live and work together under the rules of their religion.

republic (page 162) A republic is a state or nation that is not ruled by a monarch.

reservoir (page 201) A reservoir is a place where water is stored for future use.

revolution (page 106) A revolution is a dramatic, sudden change.

robot (page 347) A robot is a machine that looks like a person and can do some of the jobs people do.

ruins (page 133) Ruins are the damaged remains of very old human-made structures.

rural (page 49) Rural areas are places that are in the countryside, not near a city.

Russian Orthodox Church (page 161) The Russian Orthodox Church is the religion of many European Russians.

Russian Revolution (page 145) The Russian Revolution was the revolt in 1917 of the Russian people against the czar when Communists won control.

rye (page 152) Rye is a grain that grows well in a cool climate. It is used to make a dark bread.

sacred (page 295) Sacred means holy.

Sahel (page 194) The Sahel is a region of dry grasslands south of the Sahara. The Sahel stretches from the Atlantic Ocean to the Red Sea.

sand dune (page 273) A sand dune is a hill of sand formed by the wind.

sanitation (page 321) Sanitation means keeping something like a city clean.

savanna (page 194)   A savanna is a land area with long, thick grass and short trees.

scarce (page 257)   Scarce means hard to get.

sea level (page 42)   Sea level, or the ocean surface, is the zero point for measuring the elevation of land.

secular (page 266)   Secular means not religious.

semiarid (page 194)   A semiarid climate is hot, with rainy and dry seasons.

service job (page 18)   A service job is a job, such as teacher, doctor, clerk, or waiter, where people help other people.

shah (page 283)   Shah was the title of the king of Iran.

shifting agriculture (page 291) Shifting agriculture is a type of farming in which people move from place to place chopping down trees and planting crops until the land is no longer fertile.

Shinto (page 346)   Shinto is a religion practiced in Japan. People who believe in the Shinto religion pray to gods found in rivers, mountains, trees, and other forms of nature.

shish kebab (page 267)   Shish kebab is a popular Turkish dish of meat and vegetables cooked on a stick.

shortage (page 163)   A shortage exists when there is not enough of something, such as food or houses, for all the people who want it.

Sikhism (page 297)   Sikhism is an Indian religion.

silt (page 249)   Silt is the tiny pieces of soil left behind by a river.

sisal (page 210)   Sisal is a plant used to make rope.

slum (page 50)   A slum is an area in or around a city where poor people live crowded together in small, poorly made homes.

social problem (page 163)   A social problem is a problem, such as illiteracy, poverty, crime, or pollution, that is caused by and affects people.

software (page 258)   Software is the information and programs used by a computer.

solar energy (page 139)   Solar energy uses energy from the sun to do work.

Solidarity (page 153)   Solidarity was an organization of Polish trade unions. Now it is a political party led by Lech Walesa.

Southern Hemisphere (page 41) The Southern Hemisphere is the half of Earth south of the Equator with Australia, Antarctica, and parts of Africa and South America.

sparsely populated (page 362) A sparsely populated area has very few people living there.

standard of living (page 11) The standard of living measures how well people in a region live.

station (page 363)   A station is a large sheep or cattle ranch in Australia.

steel (page 100)   Steel is a strong metal made from iron.

steppe climate (page 242)   A steppe climate gets just enough rain for grasses to grow.

steppes (page 160)   The steppes are grassy plains. Steppes usually have fertile farmland.

stilt (page 315)   A stilt is a pole used to support a house built up off the ground.

strait (page 265)   A strait is a narrow body of water that connects two larger bodies of water.

strike (page 155)   In a strike, workers refuse to do their jobs until they get what they want, such as more money or better working conditions.

subcontinent (page 290)
A subcontinent is a large area of land that is smaller than a continent.

subsistence farmer (page 43)
A subsistence farmer works on a small farm, growing just enough food for the farmer's own family with nothing left over to sell.

suburb (page 11)   A suburb is a neighborhood close to a city.

Swahili (page 210)   Swahili is the official language of Kenya.

swamp (page 200)   A swamp is soft, wet land.

tariff (page 34)   A tariff is a tax that is paid on imported goods, making them more expensive.

technology (page 18)   Technology is new knowledge, inventions, tools, and skills to help people live or work better.

temperate zone (page 330)
The temperate zones are the climates north and south of the tropics. The weather in a temperate zone is not too hot or too cold.

terrace (page 347)   A terrace is a large, flat area of land built into the side of a hill for planting crops.

territory (page 24)   A territory is a place ruled by another nation, but it is not a full part of that nation.

terrorism (page 83)   Terrorism is the use of dangerous acts against the people of a country.

terrorist (page 83)   A terrorist is a person who uses dangerous acts against the people of a country.

Tet Festival (page 306)   The Tet Festival is the celebration of the Vietnamese New Year.

textile (page 307)   A textile is a kind of cloth.

theme (page 2)   A theme is an important main idea. There are five themes of geography: location, place, human/environment interaction, movement, and region.

timber (page 289)   Timber is wood used to make buildings and furniture.

tortillas (page 49)   A tortilla is a flat bread made from corn or wheat.

tourism (page 65)   Tourism is an industry that serves the needs of visitors, such as in hotels, restaurants, and stores.

tourist (page 47)   A tourist is a person who visits a place for pleasure and fun.

trade union (page 153)   A trade union helps workers get better salaries and better working conditions.

traditional (page 49)   Traditional means doing something the same way it has been done in the past.

traditional method (page 196) A traditional method is the old way of doing things.

tribe (page 195)   A tribe is a group of people who share a language, religion, and culture.

tropical climate (page 41)   A tropical climate is a climate near the Equator. It is hot all the time with a lot of rain.

tropical rain forest (page 42) A tropical rain forest is a thick forest that grows near the Equator where the climate is very hot and very wet.

tropical storm (page 65)   A tropical storm is a type of storm with heavy rains and strong winds that occurs in the tropics.

tropics (page 41)   The tropics is a region located near the Equator. The tropics has a very hot, rainy tropical climate.

tsetse fly (page 234)   A tsetse fly is an African insect that causes "sleeping sickness."

tsunami (page 331)   A tsunami is a huge, dangerous wave caused by an underwater earthquake.

tundra (page 160)   The tundra is far northern plains with cold, icy soil covered with permafrost.

two-way radio (page 363)   A two-way radio allows people to talk to each other.

typhoon (page 305)   A typhoon is a dangerous tropical Asian storm.

unemployment (page 116) Unemployment means people do not have jobs.

unfavorable balance of trade (page 252)   An unfavorable balance of trade means a country buys more goods from other countries than it exports.

unguarded border (page 12) There are no soldiers along an unguarded border.

unification (page 123)   Unification is the act of joining separate parts into one country.

uninhabited (page 369)   Uninhabited means without people.

United Nations (page 234)   The UN is an organization of countries that work together.

university (page 116)   A university is a type of school that people can go to after they finish high school.

untouchable (page 297) An untouchable is a person who belongs to the lowest group of people in the caste system of India.

upper class (page 50)   The upper class is a small group of rich people who own most of the land, businesses, and money.

uranium (page 114)   Uranium is a mineral used to make nuclear energy.

urban (page 11)   An urban area is in the city or suburbs, not in the country.

urbanization (page 203)   Urbanization is the movement of people to cities.

variety (page 15)   Variety means many different kinds, usually at the same time or place.

vegetarian (page 297)   A vegetarian is a person who does not eat meat.

vegetation (page 193)   Vegetation is the plants that grow in a region.

Viet Cong (page 306)   During the Vietnam War, the Viet Cong were Communists who lived in South Vietnam.

volcanic ash (page 313)   Volcanic ash is the small pieces of rocks that come out of an erupting volcano.

waterway (page 26)   A waterway is a place, such as a river or a canal, where ships can travel carrying people or materials.

Western Hemisphere (page 67)   The Western Hemisphere is the half of Earth with North and South America.

wet rice farming (page 292)   Wet rice farming is a type of farming in which rice seeds are planted in small flooded fields after heavy rains.

wildlife (page 211)   Wildlife is all the wild animals that live in an area.

world power (page 107)   A world power is a nation that is an important leader among all countries.

yam (page 202)   A yam is a vegetable similar to a sweet potato.

INDEX